Wissenschaftliche Reihe Fahrzeugtechnik Universität Stuttgart

Reihe herausgegeben von

André Casal Kulzer , Stuttgart, Deutschland

Hans-Christian Reuss, Stuttgart, Deutschland

Andreas Wagner, Stuttgart, Deutschland

AF385735

Das Institut für Fahrzeugtechnik Stuttgart (IFS) an der Universität Stuttgart forscht interdisziplinär sowie technologieoffen an modernen und zukunftsorientierten Fahrzeugkonzepten. In enger Zusammenarbeit mit Partnern aus Industrie und Wissenschaft entstehen neue Lösungen für die Mobilität der Zukunft. Das Institut gliedert sich in drei spezialisierte Lehrstühle, die gemeinsam das gesamte Spektrum der Fahrzeugtechnik abdecken: Der **Lehrstuhl für Fahrzeugantriebssysteme** widmet sich der Forschung nachhaltiger Antriebslösungen für künftige Mobilitätskonzepte. Im Fokus stehen alternative, elektrische sowie hybride Antriebssysteme und deren Komponenten – einschließlich der Nutzung nachhaltiger Energieträger wie Wasserstoff, synthetischer Kraftstoffe und Batterien. Der **Lehrstuhl für Kraftfahrzeugmechatronik** beschäftigt sich mit vernetzten, intelligenten und adaptiven Fahrzeugen. Im Zentrum stehen Fragestellungen zum Automatisierten und Vernetzten Fahren, Diagnose, Ladetechnologien, verteilte Systeme sowie softwarebasierte Fahrzeugfunktionen. Der **Lehrstuhl für Kraftfahrwesen** erforscht die physikalischen Grundlagen der Auslegung zukünftiger Fahrzeugkonzepte. Im Mittelpunkt stehen die Bereiche Aerodynamik, Windkanaltechnik, Akustik/NVH, Fahrzeugdynamik, Reifenmanagement und Thermomanagement Gesamtfahrzeug. Das IFS verfügt über eine vielfältige und hochmoderne Forschungsinfrastruktur, die realitätsnahe Untersuchungen vom Einzelbauteil bis zum Gesamtfahrzeug ermöglicht. Besonders hervorzuheben sind der Multikonfigurations- und Antriebsprüfstand für komplexe Antriebskonzepte, der Stuttgarter Fahrsimulator zur Untersuchung menschlichen Fahrverhaltens, der Aeroakustik-Fahrzeugwindkanal für akustische und strömungstechnische Fragestellungen sowie der Thermowindkanal zur Analyse thermischer Prozesse im Gesamtfahrzeug. Die wissenschaftliche Reihe „Fahrzeugtechnik Universität Stuttgart" dokumentiert die im Rahmen von Promotionen am IFS entstandene Beiträge zur Mobilität der Zukunft und zeigt deren thematische Vielfalt, methodische Tiefe und Praxisrelevanz.

Reihe herausgegeben von

Prof. Dr.-Ing. André Casal Kulzer
Lehrstuhl für Fahrzeugantriebssysteme
Institut für Fahrzeugtechnik Stuttgart
Universität Stuttgart
Stuttgart, Deutschland

Prof Dr.-Ing. Andreas Wagner
Lehrstuhl für Kraftfahrwesen
Institut für Fahrzeugtechnik Stuttgart
Universität Stuttgart
Stuttgart, Deutschland

Prof. Dr.-Ing. Hans-Christian Reuss
Lehrstuhl für
Kraftfahrzeugmechatronik
Institut für Fahrzeugtechnik Stuttgart
Universität Stuttgart
Stuttgart, Deutschland

Xiao Fei

The Impact of Unsteady Flow on Drag Measurements in Automotive Wind Tunnels

Xiao Fei
IFS, Chair of Automotive Engineering
University of Stuttgart
Stuttgart, Germany

Zugl.: Dissertation Universität Stuttgart, 2025
D93

ISSN 2567-0042   ISSN 2567-0352   (electronic)
Wissenschaftliche Reihe Fahrzeugtechnik Universität Stuttgart
ISBN 978-3-658-51765-6   ISBN 978-3-658-51766-3   (eBook)
https://doi.org/10.1007/978-3-658-51766-3

© The Editor(s) (if applicable) and The Author(s), under exclusive license to Springer
Fachmedien Wiesbaden GmbH, part of Springer Nature 2026

This work is subject to copyright. All rights are solely and exclusively licensed by the Publisher,
whether the whole or part of the material is concerned, specifically the rights of translation, reprint-
ing, reuse of illustrations, recitation, broadcasting, reproduction on microfilms or in any other
physical way, and transmission or information storage and retrieval, electronic adaptation, computer
software, or by similar or dissimilar methodology now known or hereafter developed.
The use of general descriptive names, registered names, trademarks, service marks, etc. in this
publication does not imply, even in the absence of a specific statement, that such names are exempt
from the relevant protective laws and regulations and therefore free for general use.
The publisher, the authors and the editors are safe to assume that the advice and information in this
book are believed to be true and accurate at the date of publication. Neither the publisher nor the
authors or the editors give a warranty, expressed or implied, with respect to the material contained
herein or for any errors or omissions that may have been made. The publisher remains neutral with
regard to jurisdictional claims in published maps and institutional affiliations.

This Springer Vieweg imprint is published by the registered company Springer Fachmedien
Wiesbaden GmbH, part of Springer Nature.
The registered company address is: Abraham-Lincoln-Str. 46, 65189 Wiesbaden, Germany

If disposing of this product, please recycle the paper.

Foreword / Acknowledgements

This body of work, including this dissertation, was accomplished during my time as a Research Associate at the Institute of Automotive Engineering (IFS) of the University of Stuttgart.

I would like to thank Prof. Dr.-Ing. Andreas Wagner and Prof. Dr.-Ing Jochen Wiedemann for the supervision of my work, their time invested in fruitful discussions and their valuable advice. I am also grateful for Hon.-Prof. Dr.-Ing, habil. Thomas Schütz for his efforts as co-examiner as well as his suggestions and ideas.

Furthermore, I would like to thank Dr.-Ings. Timo Kuthada and Felix Wittmeier for their continued support throughout the years working on my research. Their help both in scientific as well as organizational matters were invaluable. In addition, I would like to thank all my colleagues in Aerodynamics and Thermal Management, past and present, for an outstanding scientific and social work environment. In particular, I would like to mention Dr.-Ings. Daniel Stoll and Christoph Jessing, without whose previous contributions on unsteady flow and personal support this work would not have been possible.

I would like to give a special shout-out to my colleagues working in the Model Scale Wind Tunnel, especially Messrs. Stefan Schmidt and Bernd Geiger, who endured my, at times, haphazard experimental ideas and last-minute changes in test plans. Without them the breadth of results in this work would not have been achieved.

Finally, I would like to thank my family and friends for their support and for being with me all these years. A final thank you goes out to my partner, Erika, who had the misfortune to listen to my shoddy presentation drafts and suffer my angry rants.

Xiao Fei

Table of Contents

List of Figures

List of Tables

List of Abbreviations

BEV	Battery-Electric Vehicle
CFD	Computational Fluid Dynamics
DES	Detached Eddy Simulation
DLR	German Aerospace Center
DMWK	(IFS) Digital Model Scale Wind Tunnel
DWT	(IFS) Digital Wind Tunnel
EADE	European Aerodynamic Data Exchange
FFT	Fast Fourier Transform
FKFS	Research Institute for Automotive Engineering and Powertrains Stuttgart
ICE	Internal Combustion Engine
IFS	Institute of Automotive Engineering Stuttgart
"L"	Large Stagnation Body (76 mm extension)
MWK	(IFS) Model Scale Wind Tunnel
"none"	No Stagnation Body
NWG	(Toyota) Natural Wind Generator
RSS	Residual Sum of Squares
"S"	Small Stagnation Body (48 mm extension)
SAE	Society of Automotive Engineers
SB	Stagnation Body
swing	(FKFS) Side Wind Generator
TFI	Turbulent Flow Instrumentation
TLS	Turbulent Length Scale
Tu	Turbulence Intensity
VR	(PowerFLOW) Variable Resolution
WLTP	Worldwide Harmonized Light Vehicle Test Procedure
WT	Wind Tunnel

Abstract

Contemporary automotive aerodynamic development occurs mainly in steady-state environments, both in the wind tunnel and with numerical flow simulations. However, this does not represent actual on-road drag well, as the on-road flow state is characterized by unsteady turbulence. The goals of this work were therefore to improve the understanding of both the effects of unsteady flow on passenger cars as well as the workings of an active gust generation system in an automotive open-jet wind tunnel. Even though many investigations into unsteady flow have been made in an automotive context, the existing literature still cannot answer all questions concerning unsteady drag. First of all, the existing studies each only cover a very limited number of geometry and/or flow signal variations. Secondly, they rarely discuss how the simulation environment, be it wind tunnel geometry of CFD setup, influences the results. These points were thus the main focus of this work.

Wind tunnel tests were carried out in the University of Stuttgart Model Scale Wind Tunnel, with the FKFS swing active gust generation system installed. FKFS swing consists of 6 rotating vertical airfoils placed at the wind tunnel nozzle exit, capable of creating customized, time-resolved yaw angle signals. 16 filtered noise signals and 20 sine signals were evaluated. The noise signals offer a gaussian flow angle distribution and thus resemble realistic on-road signals. Compared to the noise signal, the sine signals' flow angle distribution does not resemble those on the road. However, they form a good baseline for scientific investigation due to their simplicity.

Measurements undertaken in the empty wind tunnel test section revealed that unsteady flow signals generated with a turbulence generation system create a static pressure distribution that is both different from the steady-state pressure distribution as well as unique to each flow signal. A non-zero static pressure gradient in a wind tunnel creates a net force on a vehicle that does not exist on the road. Ordinarily, this is not a problem, because this net force would be the same or similar for each test, which would not preclude aerodynamic optimization of the test vehicle. However, when unsteady flow is applied and the pressure gradient thus changed, the subsequent change in drag the vehicle experiences will stem from both the unsteady flow effects on the vehicle and the change in pressure gradient. In order to isolate the former, interference corrections become necessary for any drag measurements with active gust generation systems applied. Mercker and Cooper's "Two-Measurement Correction" is the contemporary method for open-jet wind tunnel interference corrections. It iterates the final corrected drag value using two measurements undertaken with different static pressure gradients in the same test section. In the Model Scale Wind Tunnel, a distinct second pressure gradient can be created by using stagnation bodies in the collector to increase the pressure behind the model, enabling the correction.

The correction method needs to be evaluated for unsteady test cases, because it has only been validated for steady-state application previously. Typically, interference corrections have been validated by applying them to the same vehicle configuration in different wind tunnels. If the corrected drag coefficient deviated by around 0.002 or less, it would mean that the corrected drag is independent of the wind tunnel geometry, thus validating the method. In order to enable this validation in the Model Scale Wind Tunnel, two differently sized stagnation bodies ("S" and "L") could be installed in the collector. Together with the standard configuration without stagnation bodies ("nSB"), this created 3 distinct permutations the correction could be applied on ("nSB" with "S"; "nSB" with "L" and "S" with "L"), simulating different wind tunnel geometries. It was found that the permutation "S" with "L" did not result in satisfactory values compared to the other two, because they created extremely high pressures near the collector. However, the other two permutations largely fulfilled the 0.002 delta criterium, and thus the average corrected drag value from these permutations was chosen for all further evaluations.

Four different quarter scale vehicle models were evaluated, using both notchback and squareback rear ends: The SAE reference body, the SAE Type K model, the DrivAer, and the AeroSUV. These models and their variations allowed the isolation of different geometric characteristics and the evaluation of said characteristics under unsteady flow.

The measurements have shown that the influence of unsteady flow on drag has two components, namely the yaw angle magnitude and the frequency profile. The yaw angle influence is similar to the relationship of steady-state yaw to drag. For every vehicle model, the drag coefficient increased with increasing signal standard deviation, just as it did with an increasing steady-state yaw angle. However, the frequency influence was unique to the unsteady flow and impossible to evaluate using steady-state methods. To assess these frequency effects, two quantifiers were introduced: Drag Sensitivity (S_D) and Average Delta to Quasi-Steady ($\Delta_{QS,avg}$).

Drag Sensitivity is calculated by forming a second-order polynomial fitting curve for the whole data set of a single vehicle (drag over signal standard deviation), determining the residual sum of squares for the data points, and then calculating the square root of the residual sum divided by the number of data points. Thus, "sensitivity" represents a weighted average offset of the data set's drag values from the fitting curve. Average Delta to Quasi-Steady is calculated in a similar way to Drag Sensitivity, the difference being that it is not the delta to a fitting curve being evaluated but the delta of each measurement point to their quasi-steady equivalent. The quasi-steady drag for each vehicle and signal combination is calculated by creating a weighted average drag out of steady-state yaw measurements, using the flow angle probabilities in the unsteady signal as weight coefficients. Consequently, high S_D means that the vehicle in question experiences big changes in drag when the signal frequency profile is changed, while high $\Delta_{QS,avg}$ means that the vehicle's unsteady drag cannot be accurately estimated using steady-state methods.

Using the aforementioned 36 distinct signals, Drag Sensitivity and Average Delta to Quasi-Steady values were calculated for all vehicle models in order to determine the geometric traits that affect these quantifiers. It was found that increases in magnitude of unsteady effects on drag were largely due to rear end interactions. Both Drag Sensitivity and Average Delta to Quasi-Steady were increased with increasing separation area (e.g. change from notchback to squareback or DrivAer to AeroSUV) or with the presence of the SAE body-like sharp-edged rear geometry.

Subsequently, CFD simulations were carried out to understand the flow effects that contribute to drag changes in unsteady flow and to evaluate the experimental accuracy of the FKFS swing gust generation system. The simulations were carried out in Simulia PowerFLOW, a Lattice-Boltzmann code. The wind tunnel measurements have previously established that the influence of signal frequency on drag represents the true unsteady influence, as opposed to the amplitude influence which can be recreated by steady-state yaw. Therefore, the simulations focused on sine signals with the same amplitude of 5° but different frequencies. For the vehicle model, the DrivAer squareback model was chosen. As a well-known, realistic reference model that has been utilized in many publications, the DrivAer's simulation models are well validated for both steady and unsteady flow.

The model was simulated in two different simulation environments: The simulated wind tunnel ("Sim. WT"), and the simulated free flow environment, ("Sim. Free Flow"). The simulated wind tunnel is a recreation of the physical Model Scale Wind Tunnel, including the nozzle, collector, test section and plenum geometries. The unsteady incident flow is created and controlled by a geometrical reproduction of the swing system. The free flow environment is a typical box environment with slight adaptations to account for unsteady flow. The unsteady incident flow is created and controlled by time-resolved inlet boundary conditions.

The simulations have shown that the changes in drag the vehicle experiences when signal frequency is changed are mostly due to changes in wake shapes. When low-frequency unsteady flow was introduced, the vehicle wake became larger and more pointed, thereby increasing drag. However, with further increasing signal frequency at constant signal amplitude, the wake slowly reverted back towards the steady-state shape, becoming smaller and more bulbous, which in turn decreased drag. In conclusion, the largest effects of unsteady flow on drag occur at low frequencies, also owing to the fact that the resulting large eddy sizes contain a large amount of energy.

The drag coefficients resulting from CFD showed that while the experiment and simulated wind tunnel show a decent match, there is a large difference to the simulated free flow environment when unsteady flow is applied. Drag differences were mainly evident at the front and rear of the vehicle and near the wheels. These areas were subsequently investigated in detail, with turbulence values and flow angles evaluated using virtual probes. At the front of the vehicle, the flow angles the vehicle experienced was observed to be different between the Sim. WT and the Sim. Free Flow, despite the flow angle time histories being identical without a vehicle in the two simulation environments. The introduction of a vehicle into the test section evidently changes the incident flow depending on the exact vehicle geometry. Since higher flow angles mean higher drag, this explained a part of the drag difference between the two environments. Near the front wheels of the vehicle, a number of differences between the wind tunnel and free flow environments were observed. Turbulence values developed in different directions with increasing signal frequency. For instance, with increasing frequency the turbulence intensity near the front wheels increased in the wind tunnel but decreased in free flow. At the rear of the vehicle, however, both wake shapes and turbulence values were largely in agreement between wind tunnel and free flow. The development of turbulence parameters with increasing flow frequency were the same between the two simulation environments. As the rear end is the most relevant part of a road vehicle with respect to aerodynamic drag, this was a positive result for the correlation between wind tunnel and real-world flow. The aforementioned differences at the rear can be explained by the static pressure gradient created by the unsteady incident flow in the wind tunnel, whose effect can be eliminated using interference corrections.

Further simulations without a vehicle in the test section / the free flow environment have discovered the cause of the discrepancies between wind tunnel and free flow near the front wheels by evaluating z-vorticity values on the left and right sides of the freestream. While in the free flow environment z-vorticity was low magnitude and in phase between the left and right sides, in the wind tunnel z-vorticity magnitude was significantly higher and offset by half a phase between the left and right sides. These observations showed the so-called "flutter" coherent flow phenomenon to exist in the wind tunnel when the gust generation system is activated. Such jet instability phenomena have been observed even without gust generation systems in a number of wind tunnels. However, jet instability can usually be suppressed through various measures near the wind tunnel nozzle. When a gust generation system like FKFS swing is activated, however, the sine signal re-excites the flow and therefore recreates the jet instability. The higher turbulence values in the wind tunnel with increasing frequency can be explained by the lower wavelength that corresponds to the higher frequency, causing more z-vortices to affect the vehicle at any point in time compared to the low frequency case.

Having discovered these shortcomings of an active gust generation system, one must set reasonable limits for the use of the swing system so that the results remain useful for correlation between experiment / CFD and reality. Realistic on-road signals have a gaussian flow angle distribution and many different discrete frequencies in one signal. The latter will already mitigate some of the aforementioned problems, since this makes it less likely that single frequencies will be able to cause compounding jet instabilities. Furthermore, the flow angle distribution being centered around $0°$ means that the average yaw angle on the road, especially at high speeds, is not particularly large. Additional CFD simulations undertaken with $2.5°$ sine amplitude instead of $5°$ have subsequently shown that at smaller amplitudes the inconsistencies between wind tunnel and free flow become sufficiently small. Consequently, this work shows that FKFS swing is well suited to evaluate drag under typical on-road flow regimes, which rarely exceed yaw angle distributions greater than the equivalent of a 2.5° sine flow.

All in all, the presented research results constitute an important contribution to understanding unsteady flow effects on drag and the aerodynamic phenomena created by an active gust generation system in an open-jet wind tunnel. Furthermore, it can serve as a new starting point for both further research on the topic and a potential robust process for vehicle aerodynamic development in realistic, unsteady flow environments.

Kurzfassung

Heutzutage wird die Entwicklung der Fahrzeugaerodynamik, sowohl im Windkanal als auch in der numerischen Simulation, gemeinhin in Umgebungen mit stationärer Anströmung durchgeführt. Allerdings resultiert dies in Abweichungen im Vergleich zum tatsächlichen Luftwiderstand auf der Straße, da die instationäre Turbulenz bei der Straßenfahrt hierbei nicht berücksichtigt wird. Die Ziele dieser Arbeit sind somit, einerseits die Effekte der instationären Anströmung auf Pkw und andererseits die strömungsmechanischen Mechanismen eines aktiven Turbulenzgenerators in einem Automobilwindkanal mit offener Messstrecke besser zu verstehen. Obwohl in der Vergangenheit bereits viele Untersuchungen bezüglich der instationären Anströmung im Kontext der Pkw-Aerodynamik durchgeführt wurden, ist die vorhandene Literatur nicht in der Lage, alle Fragen zum instationären Widerstand zu beantworten. Erstens behandeln die existierenden Studien nur eine sehr limitierte Anzahl von Geometrie- und/oder Signalvariationen. Zweitens wird der Einfluss der Simulationsumgebung auf die Ergebnisse, sei es der Windkanalgeometrie oder der CFD-Simulationsaufbau, selten thematisiert. Diese beiden Punkte sind daher der Hauptfokus der vorliegenden Arbeit.

Im Modellwindkanal der Universität Stuttgart wurden Windkanalmessungen mit dem eingebauten aktiven Turbulenzgenerator „FKFS swing" durchgeführt. FKFS swing besteht aus 6 am Düsenaustritt platzierten, drehbaren, vertikalen Flügeln, die individuelle, zeitvariante Schiebewinkelsignale generieren können. Insgesamt wurden 16 gefilterte Rauschsignale sowie 20 Sinussignale ausgewertet. Die Rauschsignale besitzen eine normalverteilte Winkelverteilung und ähneln dadurch der Strömung auf der Straße. Im Vergleich zum Rauschsignal entspricht zwar die Winkelverteilung der Sinussignale nicht dem auf der Straße, aber sie sind durch ihre Simplizität gut für eine wissenschaftliche Untersuchung geeignet.

Messungen in der leeren Messstrecke haben ergeben, dass durch aktive Turbulenzsysteme erzeugte instationäre Strömungssignale einen Druckgradienten verursachen, der sich einerseits vom statischen Druckgradienten unterscheidet und andererseits charakteristisch für jedes individuelle Signal ist. Ein Druckgradient im Windkanal führt zu einer resultierenden Kraft am Messfahrzeug, die in der Realität nicht existiert. Üblicherweise ist das kein Problem, da diese

Kraft bei jeder Messung ähnlich groß wäre und dadurch die aerodynamische Optimierung des Fahrzeugs weiterhin möglich wäre. Wenn jedoch die Anströmung instationär wird und sich dadurch der Druckgradient verändert, resultiert die Änderung des Luftwiderstandes am Prüfling sowohl von der instationären Strömungseffekte am Fahrzeug als auch von der Änderung des Druckgradienten. Um Ersteres zu isolieren, sind daher Interferenzkorrekturen für jegliche Widerstandsmessungen mit Turbulenzsystem notwendig. Der Stand der Technik der Interferenzkorrekturen bei offener Messstrecke ist die „Zwei-Messungen-Korrektur" von Mercker und Cooper, die einen finalen Widerstandswert aus zwei Messungen mit unterschiedlichen Druckgradienten in derselben Messstrecke iteriert. Im Modellwindkanal kann ein zweiter Druckgradient erreicht werden, indem Stauhölzer im Kollektor platziert werden. Diese erhöhen den Druck in Kollektornähe und ermöglichen damit die Korrektur.

Die Korrekturmethodik muss speziell für instationäre Anwendungsfälle untersucht werden, da die Methode bis dato nur für stationäre Anwendungen validiert wurde. Typischerweise werden Interfererenzkorrekturen bewertet, indem sie für dasselbe Fahrzeug mit derselben Konfiguration in verschiedenen Windkanälen angewandt wird. Wenn der korrigierte c_w-Wert zwischen den Windkanälen um 0.002 oder weniger abweicht, ist dieser nicht abhängig von der Windkanalgeometrie und damit die Korrekturmethodik validiert. Um dies im Modellwindkanal umzusetzen, gibt es die Möglichkeit, zwei Stauhölzer unterschiedlicher Größen („S" und „L") in den Kollektor einzusetzen. Gemeinsam mit der Standard-Konfiguration ohne Stauhölzer („nSB") wurden dadurch 3 Permutationen kreiert, die verschiedene Windkanalgeometrien simulieren und mit denen die Korrektur durchgeführt werden kann („nSB" mit „S"; „nSB" mit „L" und „S" mit „L"). Die Permutation „S" mit „L" zeigte dabei große Ausschläge im Vergleich mit den beiden anderen, weil sie teilweise extrem hohe Drücke nahe dem Kollektor erzeugten. Die anderen beiden Permutationen jedoch erreichten im Vergleich untereinander weitestgehend das 0.002 Delta Kriterium. Daher wurde der gemittelte, korrigierte Luftwiderstandbeiwert aus diesen Permutationen für alle weiteren Auswertungen angewandt.

Insgesamt wurden vier unterschiedliche 1:4 Fahrzeugmodelle untersucht, jeweils mit einer Stufenheck- und einer Vollheckkonfiguration: Der SAE Referenzkörper, das SAE Typ K Modell, der DrivAer und der AeroSUV. Diese Modelle erlauben mit ihren Variationen die Isolierung von verschiedenen geometrischen Eigenschaften und damit die Betrachtung dieser im Kontext der instationären Anströmung.

Die Windkanalmessungen haben gezeigt, dass der Einfluss der instationären Anströmung auf den Luftwiderstand sich auf zwei Komponenten aufteilen lässt, nämlich einerseits des Schiebewinkels und andererseits des Frequenzprofils. Der Schiebewinkeleinfluss ähnelte der Relation zwischen stationärem Schiebewinkel und Luftwiderstand. Für alle Prüflinge stieg der Luftwiderstand sowohl bei steigendem stationärem Schiebewinkel als auch bei steigender Standardabweichung des instationären Signals. Der Frequenzeinfluss war allerdings charakteristisch für jedes spezifische Signal und damit durch stationäre Methoden nicht auswertbar. Um diese Frequenzeffekte beurteilen zu können, wurden daher zwei Kennzahlen eingeführt: Drag Sensitivity (S_D) und Average Delta to Quasi-Steady ($\Delta_{QS,avg}$).

Drag Sensitivity wird berechnet, indem eine Ausgleichskurve zweiten Grades für das komplette Datenset eines Fahrzeuges erstellt wird (Luftwiderstand gegenüber Signal-Standardabweichung), das durchschnittliche Fehlerquadrat ermittelt wird und daraus die Quadratwurzel berechnet wird. „Sensitivity" stellt also eine gewichtete, durchschnittliche Differenz der Datenpunkte eines Fahrzeugs zu der Ausgleichskurve dar. „Average Delta to Quasi-Steady" wird ähnlich wie „Drag Sensitivity" berechnet. Hierbei wird aber nicht das Delta zu einer Ausgleichskurve, sondern das Delta zwischen jedem Messpunkt und seinem quasistationären Äquivalent berechnet. Dieser quasistationäre Widerstand wird für jede Fahrzeug-Signal-Kombination einzeln berechnet, indem ein gewichteter Durchschnitt aus stationären Widerstandsmessungen bei verschiedenen Schiebewinkeln gebildet wird, wobei die Winkelverteilung des Signals als Gewichtungsfaktor fungiert. Qualitativ bedeutet also ein hohes S_D, dass das Fahrzeug große Änderungen des c_W bei Änderungen des Signal-Frequenzprofils zu erwarten hat. Ein hohes $\Delta_{QS,avg}$ bedeutet hingegen, dass der instationäre Widerstand des Fahrzeugs nicht hinreichend mit stationären Methoden abgebildet werden kann.

Mit den zuvor erwähnten 36 Signalen wurden daraufhin Drag Sensitivity und Average Delta to Quasi-Steady für alle Fahrzeugvariationen bestimmt. Damit sollten geometrische Eigenschaften identifiziert werden, die diese Kennzahlen beeinflussen. Es wurde festgestellt, dass der Einfluss instationärer Strömung auf den Widerstand größtenteils auf Interaktionen am Fahrzeugheck zurückzuführen sind. Sowohl Drag Sensitivity als auch Average Delta to Quasi-Steady stiegen mit steigender Ablösefläche (bspw. beim Vergleich Stufenheck–Vollheck oder DrivAer–AeroSUV) oder mit dem Präsenz von einfacher, scharfkantiger Heckgeometrie, wie beim SAE Modell.

Im Anschluss an die Messungen wurden CFD-Simulationen durchgeführt, um einerseits die Effekte zu verstehen, die zu den Widerstandsänderungen unter instationärer Strömung beitragen und um andererseits die Realitätsnähe des FKFS swing Turbulenzsystems zu beurteilen. Die Simulationen wurden mit dem Lattice-Boltzmann Code Simulia PowerFLOW durgeführt. Wie zuvor erwähnt ist der Einfluss der Signalfrequenz auf den Widerstand der wahre instationäre Einfluss, weil da der Einfluss der Signalamplitude durch quasistationäre Auswertungen abgebildet werden kann. Daher fokussierten sich die Simulationen auf Sinussignale mit identischer Amplitude von 5°, aber unterschiedlichen Frequenzen. Als Fahrzeugmodell wurde das DrivAer Vollheckmodell ausgewählt. Dieses bekannte Referenzmodell ist bereits in einer Vielzahl an Publikationen erforscht worden und besitzt daher bereits hinreichend validierte Simulationsaufbauten.

Das Fahrzeugmodell wurde in zwei verschiedenen Umgebungen simuliert: Der simulierte Windkanal und das simulierte Freifeld. Der simulierte Windkanal ist eine Nachbildung des physischen Modellwindkanals, inklusive der Geometrien der Düse, des Kollektors, der Messstrecke und des Plenums. Die instationäre Anströmung wird vom geometrisch nachgebildeten swing-System erzeugt und kontrolliert. Das simulierte Freifeld ist eine typische Box-Umgebung mit wenigen Anpassungen für die instationäre Anströmung. Hier wird die instationäre Anströmung durch eine zeitvariante Randbedingung am Inlet erzeugt und kontrolliert.

Die Simulationen haben gezeigt, dass die Differenz des Luftwiderstandes bei unterschiedlichen Signalfrequenzen größtenteils Nachlaufverformungen zuzuschreiben ist. Wenn niederfrequente Sinusströmung eingebracht wurde, wurde der Fahrzeugnachlauf spitzer und länger und vergrößerte somit den Widerstand. Wenn jedoch die Frequenz weiter erhöht wurde, veränderte sich der Nachlauf wieder in Richtung der stationären Form, d.h. er wurde wieder bauchiger und kürzer, wodurch der Widerstand wieder geringer wurde. Die größten instationären Widerstandseffekte traten schlussendlich bei niedrigen Signalfrequenzen auftraten, was auch an dem hohen Energiegehalt der großskaligen Wirbel bei solchen Frequenzen liegt.

Die Luftwiderstandsbeiwerte aus der CFD zeigten zwar eine gute Korrelation zwischen Windkanalmessung und simuliertem Windkanal, aber auch eine große Differenz zwischen Windkanal und Freifeld unter instationärer Anströmung. Widerstandsunterschiede traten vor allem an Front, Heck und nahe den

Rädern des Autos auf. Diese Bereiche wurden daraufhin im Detail betrachtet, indem Turbulenzkennzahlen und Strömungswinkel mit virtuellen Sonden untersucht wurden. Hierbei wurde festgestellt, dass der Strömungswinkel, welches das Fahrzeug an der Fahrzeugfront erfährt, sich zwischen simuliertem Windkanal und Freifeld unterschied, obwohl die Zeithistorien der Strömungswinkel zwischen den zwei Umgebungen ohne Fahrzeug identisch waren. Offenbar verändert ein Fahrzeug in der Messstrecke die Anströmung, wobei diese Veränderung abhängig von der exakten Fahrzeuggeometrie ist. Da größere Strömungswinkel einen höheren Luftwiderstand erzeugen, erklärt dies einen Teil der Unterschiede zwischen den beiden Umgebungen. An den Vorderrädern des Fahrzeugs konnten einige Unterschiede zwischen dem simulierten Windkanal und Freifeld festgestellt werden. Die Änderung von Turbulenzkennzahlen mit steigender Signalfrequenz war in den beiden Umgebungen entgegengesetzt. Beispielsweise stieg die Turbulenzintensität nahe den Vorderrädern mit steigender Frequenz im Windkanal, sank aber im Freifeld. Am Fahrzeugheck jedoch waren Nachlaufform und Turbulenzwerte zwischen Windkanal und Freifeld weitestgehend deckungsgleich. Die Entwicklungen von Turbulenzkennzahlen mit steigender Signalfrequenz waren zwischen den beiden Umgebungen ebenfalls identisch. Da der Heckbereich der relevanteste Teil eines Pkws ist, was den Luftwiderstand betrifft, waren diese Resultate positiv für die Korrelation zwischen Windkanal und Straßenfahrt. Die vormals festgestellten Unterschiede am Heck, was den Widerstand betrifft, entstammen dem aus der instationären Anströmung entstandenen Druckgradienten im Windkanal, der aber mithilfe von Interferenzkorrekturen ausgeklammert werden kann.

Simulationen ohne Fahrzeug in der Messstrecke / in der Freifeldumgebung, die im Weiteren durchgeführt wurden, konnten die Differenzen zwischen Windkanal und Freifeld nahe den Vorderrädern erklären. Dazu wurde die z-Wirbelstärke links und rechts im Freistrahl in beiden Umgebungen ausgewertet. Im Freifeld war die z-Wirbelstärke relativ klein und links und rechts der Symmetrieebene gleichphasig, während im Windkanal die z-Wirbelstärke um ein Vielfaches größer und links und rechts eine halbe Phase zueinander versetzt waren. Diese Beobachtungen zeigen den Präsenz der kohärenten Struktur des „Flatterns", wenn das Turbulenzsystem aktiviert ist. Solche Strahlinstabilitäten wurden zuvor auch ohne Turbulenzsysteme in Windkanälen festgestellt, konnten aber üblicherweise durch Maßnahmen an der Düse unterbunden werden. Mit einem aktiven Turbulenzsystem wie dem FKFS swing wird die

Strömung jedoch aktiv durch das Sinussignal angeregt und die Instabilität damit wiederhergestellt. Die höheren Turbulenzwerte bei steigender Frequenz im Windkanal kann durch die geringere Wellenlänge erklärt werden, die analog zur steigenden Frequenz auftritt. Dadurch erhöht sich die Anzahl der z-Wirbel, die zu jeder spezifischen Zeit auf das Fahrzeug wirken.

Mit dem nun erlangten Wissen über die Schwächen eines aktiven Turbulenzsystems müssen darauf basierend sinnvolle Limits für das swing System gesetzt werden. Damit soll der Windkanal weiterhin realistische und aufschlussreiche Ergebnisse liefern, die zur Straßenfahrt korrelieren. Realitätsnahe Anströmung auf der Straße besitzt eine normalverteilte Schiebewinkelverteilung und viele diskrete Frequenzen in einem Signal. Letzteres wird bereits einige der erwähnten Probleme lösen, da durch die vielen Frequenzen eine starke Strahlanregung bei einer einzelnen Frequenz unwahrscheinlich wird. Darüber hinaus wird sich der Großteil der auftretenden Winkel um 0° herum befinden. Der durchschnittliche Strömungswinkel wird dadurch, insbesondere bei hohen Geschwindigkeiten, relativ klein ausfallen. Zusätzliche CFD-Simulationen mit 2.5° statt 5° Amplitude, die während dieser Arbeit durchgeführt wurden, haben daraufhin gezeigt, dass bei kleinen Amplituden die Unterschiede zwischen Windkanal und Freifeld hinreichend klein werden. Dadurch hat sich gezeigt, dass FKFS swing gut geeignet ist, den Luftwiderstand bei typischen Strömungssituationen der Straßenfahrt zu bestimmen. Dessen Winkelverteilungen sind in der Regel nicht breiter sind als die eines Sinussignals mit einer Amplitude von 2.5°.

Alles in Allem ist die hier dargelegte Forschung ein wichtiger Beitrag zum Verständnis einerseits von instationären Strömungseffekten im Zusammenhang mit Luftwiderstand, andererseits von aerodynamischen Phänomenen bei der Nutzung eines aktiven Turbulenzsystems. Weiterhin kann diese Arbeit einen neuen Startpunkt für weitere Grundlagenforschung in diese Richtung oder einen neuen, robusten Prozess für die aerodynamische Fahrzeugentwicklung in instationären, straßennahen Konditionen, bilden.

1 Introduction

Today, aerodynamic development of passenger cars is an essential part of the vehicle development process. Optimized aerodynamics offer many benefits, such as safety and stability at speed, reduced wind noise and improved visibility in adverse conditions. However, the biggest focus lies on the reduction of aerodynamic drag. Lower drag results in lower energy consumption, especially at highway speeds, where aerodynamic drag accounts for a large part of total driving resistance [1]. Lower energy consumption, consequently, directly improves the bottom line for manufacturers through mitigation of financial penalties from CO_2 emission legislation [2]. The recent surge of battery-electric vehicles (BEVs) has further amplified the importance of aerodynamic drag reduction. Due to the limited range of BEVs compared to traditional internal combustion engine (ICE) vehicles, BEV customers tend to strongly value high range. Compared to larger batteries, aerodynamic improvements are a more weight- and cost-efficient way to gain extra range. In addition, energy recovery technologies present in virtually all electric vehicles enable the partial recovery of energy spent on accelerating a larger vehicle mass, but cannot recover energy lost due to wind resistance. This further increases the importance of drag reduction [3].

In order to determine the energy consumption of a passenger car during type certification, the vehicle is ran on a test bench, with a load cycle modeled after the typical use case of such a car including city and highway driving, e.g. the WLTP cycle [2]. For this, it is necessary to enter aerodynamic drag values into the test bench to simulate the aerodynamic resistance. These drag values are often determined through measurements in wind tunnels that fulfil certain requirements set by government agencies. Typically, the requirements include traits like jet uniformity in velocity and direction, ground simulation systems, accuracy of measurement equipment and measurement reproducibility [2].

As of the writing of this thesis, all certification methods in the world require only uniform and steady-state flow. However, while such a flow field lends well to reproducibility and measurement accuracy, it does not in fact represent the real-world, on-road flow. In reality, wind, terrain and traffic all cause the flow experienced by the vehicle to be highly transient, with constantly fluctuating velocities and incident angles.

© The Author(s), under exclusive license to
Springer Fachmedien Wiesbaden GmbH, part of Springer Nature 2026
X. Fei, *The Impact of Unsteady Flow on Drag Measurements in Automotive Wind Tunnels*, Wissenschaftliche Reihe Fahrzeugtechnik Universität Stuttgart,
https://doi.org/10.1007/978-3-658-51766-3_1

To study transient flow in the wind tunnel, with the long-term aim of moving aerodynamic development closer to the on-road case, FKFS has developed the swing system, which creates a reproducible unsteady yaw signal in the wind tunnel. Swing made it possible to e.g. compare steady-state drag to unsteady drag, or reproduced flow situations measured during on-road driving in the wind tunnel. However, to date, there has been no comprehensive analysis on the effects of different, parametrized flow signals on vehicle drag. Furthermore, while the mechanics of aerodynamic drag in passenger cars are well understood, those concerning drag changes in unsteady flow are not. Finally, it remains unclear how accurate the drag measurements under unsteady flow in the wind tunnel are compared to blockage-free, turbulent on-road flow.

This thesis will provide a better understanding of unsteady flow mechanics and influences on vehicle drag both relating to the vehicle itself and to the typical automobile open-jet wind tunnel.

2 Theory Framework

A theory framework consisting of definitions, values, and other knowledge is necessary to understand the investigations and results contained in this thesis. Section 2.1 will describe the coordinate system and velocities relating to a moving vehicle as a basis for further definitions and values. Section 2.2 will introduce the basic aerodynamic values and coefficients that are important to the present research. Section 2.3 will specifically explain aerodynamic values used to characterize unsteady flow situations, including some statistical methods. Section 2.4 will describe the wind tunnel, a common aerodynamic research tool that played a big part in the present investigations. Finally, section 2.5 will summarize current numerical simulation methods in vehicle aerodynamics.

2.1 Vehicle Coordinate System and Velocities

Figure 2.1 shows the vehicle coordinate system used throughout this work. The point of origin is at the center of the vehicle front axle, i.e. at the middle of an imaginary line between the two front wheel centers. The x-axis is aligned from front to back, i.e. opposite to the direction of travel. The y-axis points right, viewed from the direction of travel, and the z-axis points up.

Whether the vehicle is driving on the road or resting inside a wind tunnel, it is always treated as a fixed reference frame. Therefore, all air velocities are given in relation to said vehicle reference frame. The velocity over ground v_g has the same magnitude of the vehicle's travel velocity, but oriented the opposite way. It represents the incident air velocity caused solely by the vehicle's movement. The wind velocity v_w represents flow velocities caused by the outside environment. This could be natural wind, like the name implies, but also be turbulences caused by nearby traffic, for instance. The vector sum of v_g and v_w is the total incident flow velocity v_∞. This is the velocity vector that simulations, either experimental or numerical, try to replicate. It represents the total incident flow the vehicle experiences.

© The Author(s), under exclusive license to
Springer Fachmedien Wiesbaden GmbH, part of Springer Nature 2026
X. Fei, *The Impact of Unsteady Flow on Drag Measurements in Automotive Wind Tunnels*, Wissenschaftliche Reihe Fahrzeugtechnik Universität Stuttgart,
https://doi.org/10.1007/978-3-658-51766-3_2

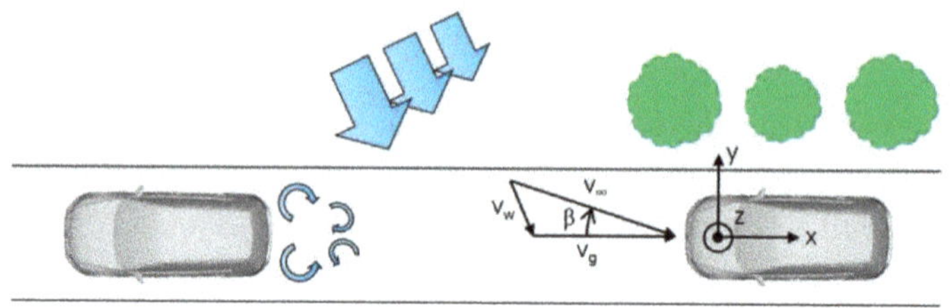

Figure 2.1:   Vehicle coordinate system and flow velocities [4]

2.2   Aerodynamic Coefficients and Dimensionless Quantities

The aerodynamic drag force F_D is a big part of a vehicle's cumulative driving resistance and the one aerodynamicists look to reduce as much as possible. However, in order to compare different vehicles and/or setups, solely comparing F_D is insufficient because drag force is not only dependent on the shape of the vehicle, but also on e.g. air speed and air density.

$$F_D = \frac{\rho}{2} v_\infty^2 c_D A_x \qquad \text{Eq. 2.1}$$

with:

ρ: air density c_D: drag coefficient
v_∞:   incident flow speed A_x: frontal area

To evaluate a vehicle's shape for aerodynamic drag, it is therefore necessary to normalize for the influences through air density, flow velocity and frontal area. This leads us to the drag coefficient c_D, which is commonly used as a comparative value in automobile aerodynamics.

$$c_D = \frac{F_D}{\frac{\rho}{2} v_\infty^2 A_x} \qquad \text{Eq. 2.2}$$

The following part of the terms in equations 2.1 and 2.2 can be defined as the incident dynamic pressure q_∞:

$$q_\infty = \frac{\rho}{2} v_\infty^2$$

Eq. 2.3

q_∞ represents the energy content of the flow contained in its movement, as a pressure value. For a low-speed environment such as in automobile applications, air density ρ can be considered independent of the velocity. In that case, dynamic pressure is simply a scalar expression for flow velocity.

Since the automobile is considered a bluff body, most of its aerodynamic drag derives from different surface pressures at different locations [5]. Like the drag force, a pressure value can also be normalized in order to facilitate comparisons. This results in the pressure coefficient c_p:

$$c_{p,i} = \frac{p_i - p_\infty}{\frac{\rho}{2} v_\infty^2}$$

Eq. 2.4

with:
p_i: static pressure at a specific location i
p_∞:   ambient pressure

Similarly, total pressure is also normalized with the incident dynamic pressure, resulting in the total pressure coefficient c_{pt}. This value represents the total energy content in the flow as a pressure value, containing static and dynamic pressure, and is a good measure to examine energy losses with.

$$c_{pt,i} = \frac{p_{t,i} - p_\infty}{\frac{\rho}{2} v_\infty^2}$$

Eq. 2.5

with:
$p_{t,i}$:   total pressure at a specific location i

In addition to the coefficients mentioned, there are further dimensionless quantities significant to the present research. These are especially important because all of the research has been carried out on scale models, where they ensure comparability with other results.

The Reynolds number Re measures the ratio between inertial and viscous forces.

$$Re = \frac{v_\infty L}{\upsilon} \qquad \text{Eq. 2.6}$$

with:

L: characteristic length

υ: kinematic viscosity

In automotive aerodynamics, the Reynolds number is usually in the magnitude of 10^7 (e.g. for a passenger car traveling at 140 km/h) due to the low viscosity of air. Therefore, inertial forces dominate [6].

L commonly refers to the vehicle or model length. Since Re is proportional to flow speed and inversely proportional to length, model scale experiments need to either increase flow speed or reduce fluid viscosity to achieve similar Re.

The Mach number M measures the ratio between flow velocity and the local speed of sound c. In practice, it limits flow velocity in model scale experiments because a Mach number above ~0.3 necessitates the consideration of fluid compressibility.

$$M = \frac{u_\infty}{c} \qquad \text{Eq. 2.7}$$

Finally, the Strouhal number St represents a dimensionless frequency value for oscillating flow.

$$St = \frac{fL}{v} \qquad \text{Eq. 2.8}$$

with:

f: oscillation frequency

As with the Reynolds number, care should be taken to match Strouhal numbers in model and full-scale flow when comparing experiments. Model scale experiments need to increase the signal frequency to counteract the reduced characteristic length and the increased flow velocity.

2.3 Unsteady Aerodynamic Quantities

Having covered basic aerodynamic values in the section 2.2, the current section will explain mathematic and aerodynamic values and principles specifically suited to unsteady evaluations.

2.3.1 Standard Deviation and Pressure Fluctuation

The standard deviation σ is a scalar that denotes the magnitude of difference to the average in a set of values. If most of the values are close to the average of the set, σ is small. The standard deviation is mathematically defined as follows:

$$\sigma_i = \sqrt{\frac{1}{n}\sum_{i=1}^{n}(x_i - \bar{x})^2} \qquad \text{Eq. 2.9}$$

with:

n: number of data points

x_i: value at a specific location i

In the context of unsteady flow, standard deviation is used to measure the fluctuation of flow values around their average and therefore to quantify specifically unsteady effects in the flow.

The pressure fluctuation $c_{p,rms}$ acts as a quantifier for unsteady behavior in pressure. Mathematically, it is the standard deviation of the pressure coefficient introduced in section 2.2.

$$c_{p,rms} = \frac{\sigma(p_i)}{\frac{\rho}{2}\overline{|v|_i^2}} * 100\ \% \qquad \text{Eq. 2.10}$$

with:

p_i: static pressure at a specific location i

v_i: flow velocity at a specific location i

2.3.2   Turbulence Intensity and Length Scale

The turbulence intensity Tu measures the velocity fluctuation in different spatial directions in relation to the velocity magnitude. Like $c_{p,rms}$, turbulence intensity is also defined with a standard deviation term.

$$Tu_{j,i} = \frac{\sigma_{j,i}}{\overline{|v|_i}} * 100\ \% \qquad \text{Eq. 2.11}$$

with:

i: specific location i

j: spatial direction (x, y or z)

$\sigma_{j,i}$ velocity standard deviation in j-direction at specific location i

v_i: flow velocity at specific location i

The turbulent length scale TLS quantifies the physical dimensions of large eddies in the flow. Multiple methods to calculate TLS exist, like the "von Kármán Length Scale" or the autocorrelation method [7] [8]. In the present work, TLS will be calculated using an autocorrelation method derived from Taylor [9]. In this method, the autocorrelation function R_{AC} compares a time-resolved signal with itself under different temporal offsets Δt.

$$R_{AC,j,i}(\Delta t) = \lim_{\tau \to \infty} \frac{1}{2\tau} \int_{-\tau}^{\tau} v_{j,i}(t) * v_{j,i}(t + \Delta t)\ dt \qquad \text{Eq. 2.12}$$

Integrating R_{AC} and subsequent multiplication with the average flow velocity magnitude results in the turbulent length scale TLS. Here, t_0 is the time where $R_{AC} = 0$ for the smallest $t > 0$.

$$TLS_{j,i} = \overline{|v|}_i * \int_0^{t_0} R_{AC,j,i}(t)\, dt \qquad \text{Eq. 2.13}$$

In equations 2.12 and 2.13 the variables are:

i: specific location i
j: spatial direction (x, y or z)
$v_{j,i}$ velocity magnitude in j-direction at specific location i

Like turbulence intensity, TLS can be calculated in all spatial directions x, y and z in order to determine the dimensions of unsteady structures in any direction.

2.3.3 Vorticity

The vorticity ω describes the local spinning motion of the flow at a specific point. While not a statistical term like pressure fluctuation, turbulence intensity and length scale, it is nonetheless a useful metric in quantifying unsteady flow phenomena concerning directional changes and vortices. It is defined as the curl of the flow velocity vector.

$$\vec{\omega} = \nabla \times \vec{u} \qquad \text{Eq. 2.14}$$

Vorticity can be specifically defined around any axis, but for the present work it is most important to consider the z-vorticity ω_z. Unsteady yawed flow is a big part of the research, which causes curls of the flow vector to be predominantly around parallels of the z-axis.

2.3.4 Fast Fourier Transformation

When evaluating unsteady flow value developments over time, it is helpful to transform the transient result collected in the time domain into the frequency domain. This allows analyses of frequency dependencies between incident and resulting flow, as well as of other characteristic frequencies at certain locations. It shows which discrete frequencies dominate a complex unsteady flow signal or result.

A common method to do so is the Fourier Transform (FT), shown in a simplified manner in **Figure 2.2**. The input signal in the time domain is treated as a sum of sine signals with discrete frequencies, which when observed from the frequency domain result in the FFT result with the discrete peaks [10].

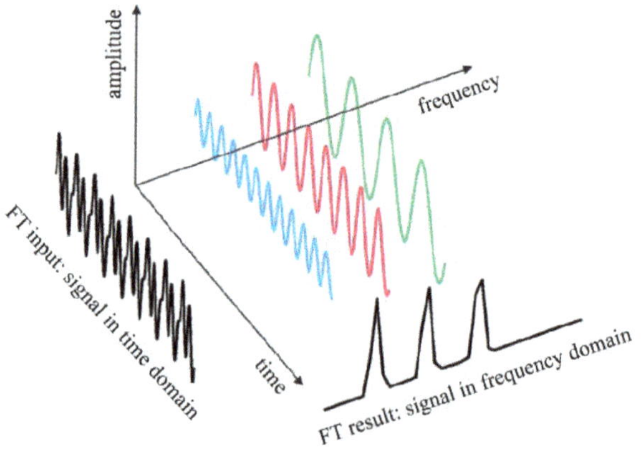

Figure 2.2:   Simplified visualization of a Fourier Transform [4]

2.4 Wind Tunnels

Automotive wind tunnels are important tools for aerodynamic development. They aim to reproduce the flow regime on the road in a controlled environment. A substantial part of the investigations in this work were conducted in the wind tunnel.

2.4.1 An Overview on Wind Tunnels

Wind tunnels are divided into two types based on their air circulation. The Eiffel type tunnel collects air from outside the wind tunnel and also releases it back to the outside, while the Göttingen type tunnel circulates the air in a closed loop. The advantages of the Eiffel type wind tunnel are mainly its lower upfront cost and space requirements and the ease of removing waste gases from the test section. However, most current automotive wind tunnels are of the Göttingen type due to its lower energy requirements and independency of outside conditions [11].

There are several different test section types in use for wind tunnels, such as the open, closed, slotted wall, streamlined wall and adaptive wall test sections. However, the vast majority of passenger car wind tunnels opt for the open test section. It provides better conditions for working on the test vehicle, gives faster turnaround times and maintains ambient pressure in the jet due to the large plenum around the test section [11].

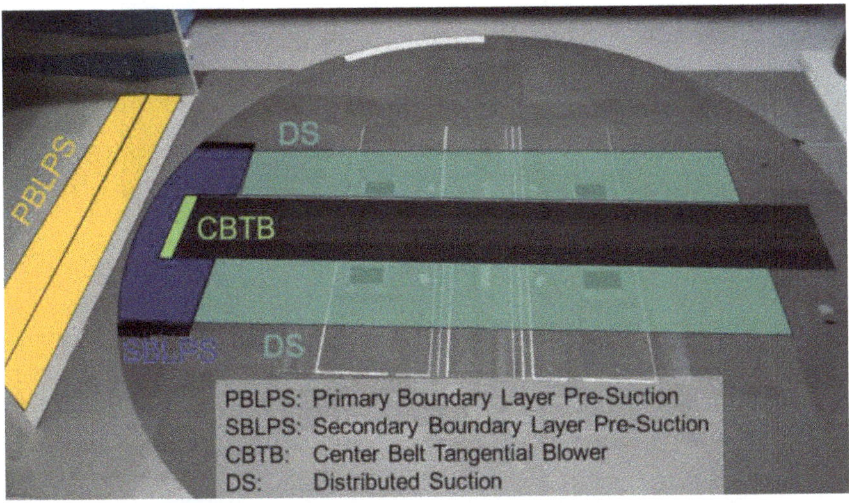

Figure 2.3:	Ground simulation systems, exemplified in the University of Stuttgart full-scale aeroacoustic wind tunnel [12]

Since the wind tunnel strives to reproduce on-road situations, it is also necessary to simulate the ground influence as well as possible. If the vehicle is stationary in the test section, the floor must move at the desired flow velocity, the wheels must rotate and the floor boundary layer created in the airline immediately before the test section must be removed. The exact ground simulation systems differ in their specifics between wind tunnels, but usually include a belt system to create a moving road and wheel rotation, as well as a boundary layer control system. Commonly, production car wind tunnels use a 5-belt system, with one center belt running between the car's wheels and four wheel drive belts rotating the wheels. The boundary layer system usually consists of some combination of: primary and secondary suction and/or scoop systems in front of the vehicle, tangential blowing in front of the belts and/or wheels, and distributed suction around the vehicle [11].

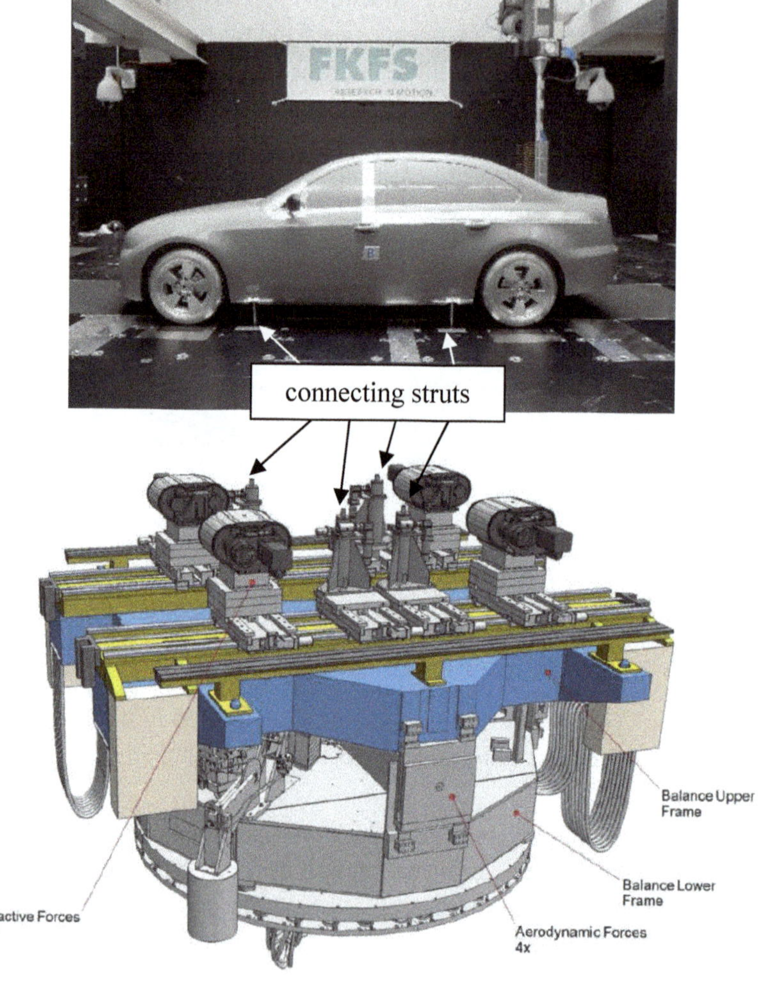

Figure 2.4:   Above: Vehicle mounted on struts in the University of Stuttgart Model Scale Wind Tunnel test section. Below: Parts of a wind tunnel underfloor balance [12]

The systems used to restrain vehicles in the test section and to measure aerodynamic forces depends on the ground simulation configuration. Usually, forces are transmitted through the ground to an underfloor balance, but in-vehicle balances for scale models also exist. In production car wind tunnels, the vehicle itself is most commonly fixated to the floor using struts connecting the balance to the underbody. Alternatives include fixation using rods or cables, or stings on top of the model. The system consisting of vehicle and balance can rotate around the z-axis to enable measurements with different, steady state incident yaw angles [11].

In order to measure flow field data, two possible devices that were also used in this work are the Prandtl probe and the multi-hole probe. The Prandtl probe measures static and total pressure as a scalar and can be used to derive further fluid values such as dynamic pressure or velocity magnitude. When it is important to track directional and/or time-resolved values, such as flow angles, the multi-hole probe can be used. Turbulent Flow Instrumentation (TFI)'s multi hole "Cobra" probe, for instance, can dynamically measure flow velocities up to 50 m/s within incident angles of $\pm45°$ and at up to 2 kHz. It is accurate to ±0.3 m/s for velocities and $\pm1°$ for flow angles [13].

Figure 2.5:   TFI's multi-hole Cobra probe [13]

In addition, flow probes are commonly mounted on multi-axis traversing systems that can automatically move through the test section and therefore automate measurements of multiple points in the airflow. Often, it is even possible to mount several probes on the traversing system at once, cutting down measurement time for big spatial grids.

2.4.2   Wind Tunnel Interference Corrections

Ideally, the drag coefficient measured in the wind tunnel is the same as the drag the vehicle experiences on the road. Realistically, this is not the case even if the road is even and free of obstacles. In an open jet wind tunnel, like the IFS Model Scale Wind Tunnel, the jet has a limited size and interacts with elements other than the vehicle, such as the nozzle and collector geometries, the plenum, and the test section floor. These interferences change the measured drag coefficient in a way that is not present on the road.

Mercker and Wiedemann were the first to develop a correction method for the open jet wind tunnel, using physical models to describe five different interference effects – four so-called "q-interference" effects and the static pressure gradient effect [14].

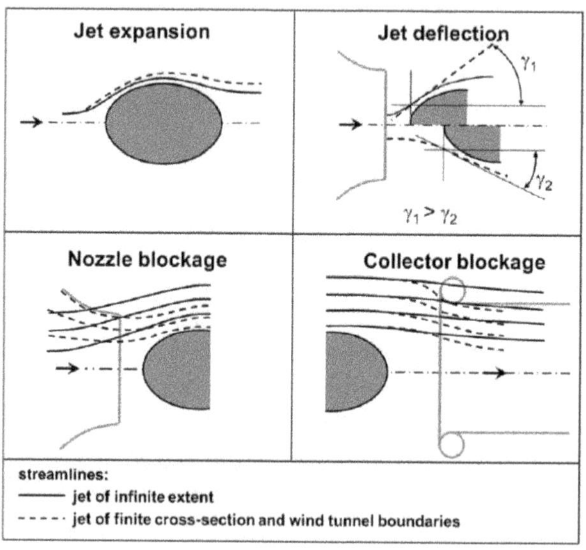

Figure 2.6:   Open jet wind tunnel interference effects affecting dynamic pressure ("q-interference") [14]

Figure 2.6 shows the "q-interference" effects, namely "jet expansion", "jet deflection", "nozzle blockage" and "collector blockage". Each of these changes the jet direction or jet shape the vehicle experiences compared to the on-road, free-flow case. The change in jet shape induces a change in flow velocity explainable by continuity laws. This change in flow velocity then changes the dynamic pressure ("q"), necessitating corrections [14].

The fifth interference effect is the static pressure gradient influence. As the static pressure along the wind axis in a wind tunnel test section is not constant, a nonzero difference in static pressure between the front and back of the vehicle will create an extra longitudinal force on the vehicle not present on the road. This force needs to be subtracted from the measured result. [14]

Subsequent development in interference corrections in an open jet wind tunnel shares Mercker and Wiedemann's methodology while fine-tuning the formulae for the different effects. The current state-of-the-art method is Mercker and Cooper's Two-Measurement correction method, which however necessitates two measurements per vehicle configuration, as the name suggests [15]. Furthermore, Hennig has added a more precise formula for collector blockage in his 2016 dissertation [16]. Lounsberry and Walter offer a summary of the current correction methodology in their work [17].

The two methods commonly used to assess the accuracy of correction methods are correlation measurements and comparisons with CFD. Typically, measuring a vehicle's drag in different wind tunnels, or in the same wind tunnel with different configurations, will result in different results due to the interference effects. However, if the corrected drag results are equal, it means that those effects have had no influence on the actual vehicle drag, validating the correction method. Mercker/Cooper [15] and Fischer [18] both used different test section configurations in the same wind tunnel to validate the correction method. One example from Fischer's work is shown in **Figure 2.7**. The difference between the largest and smallest uncorrected c_D values is ~0.014 for the different wind tunnel configurations. However, once the results are corrected, the difference between corrected c_D values reduces to ~0.001, proving the correction's accuracy.

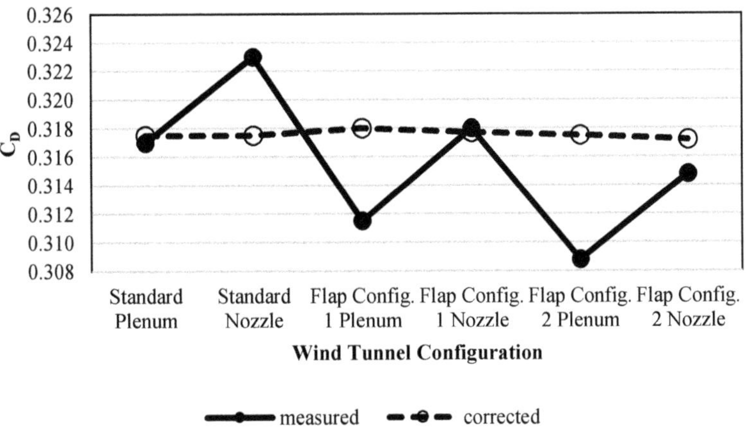

Figure 2.7: Correction results for a coupe vehicle in different wind tunnel configurations [18], replotted

Hennig on the other hand used data from the EADE 2010 correlation tests [19] where different vehicles were measured in wind tunnels across Europe and North America. The evaluation of his correction method [16] yielded similar results to Fischer, achieving a maximum c_d delta of about 0.002 for his corrected results.

In his book, Fischer also uses CFD to evaluate interference corrections. To this end, he compares the corrected measurement results with CFD results achieved from a basic, blockage-free environment. Theoretically, this blockage-free simulation represents on-road driving since there are no constraints on jet size and shape nor a static pressure gradient. The resultant deltas between corrected measurement and CFD range from 0.001 to 0.010 for different vehicles. [18]

2.5 Computational Fluid Dynamics

Aside from wind tunnels, another important tool used in automotive aerodynamics is computational fluid dynamics (CFD). Like the wind tunnel, CFD simulates on-road flow in a controlled environment but using numerical calculations instead of experimental methods. The CFD simulation process is generally divided into three steps: Pre-processing, solving and post-processing [20].

Pre-processing involves the discretization of geometries, surfaces and volumes. Usually, surfaces are broken down into planar, usually triangular pieces. Volumes are divided into small cells, often as cubes of different sizes. The smaller the sizes of the discrete surfaces and cells are, the more accurate the representation of reality becomes. However, in turn, the necessary computing time increases. In order to balance accuracy and simulation complexity, discretization is usually variable. Finer cell sizes are chosen at aerodynamically more relevant locations, like near surfaces and at expected zones of flow separation. **Figure 2.8** shows an example of variable cell resolutions on a motorcycle simulation model.

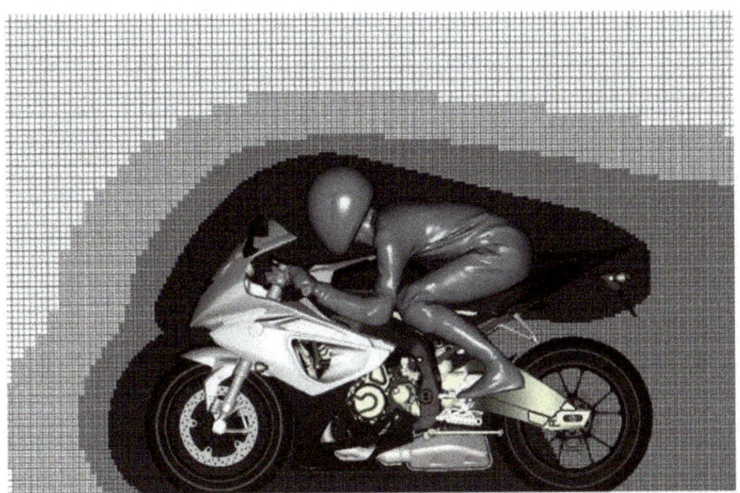

Figure 2.8: CFD volume discretization around a motorcycle model, with smaller cell sizes close to the vehicle and behind the vehicle, in its separation zone [20]

Two common solving methods for automobile aerodynamics are the Lattice-Boltzmann method (LBM) and the Navier-Stokes method. The main advantage of LBM is its inherently transient nature, which makes it well suited to investigate unsteady effects [21]. For that reason, all simulations in this work were carried out using Simulia PowerFLOW, a commercial CFD code using LBM. Aerodynamic investigations on automobiles is a common use case for PowerFLOW and have often been validated by comparing simulation results to wind tunnel measurements. Some examples can be found during the development of generic research models [22] [23], the development of production car external aerodynamics [24] and cooling drag investigations [25].

The other common fluid simulation method for automobiles is the Navier-Stokes method, in which continuity equations are solved iteratively for every discrete point in the simulation domain until equilibrium is reached. Simcenter STAR-CCM+ and OpenFOAM are two well-known CFD codes using this method. Navier-Stokes codes inherently return steady-state results and need further modifications for unsteady use cases, which is however possible.

However, due to the complexity of turbulent flow, neither LBM nor Navier-Stokes methods are able to compute it in a timely manner [20]. Therefore, turbulence models are necessary to simplify the turbulent flow effects while still offering acceptable correlation to the real world. PowerFLOW, for instance, models turbulent effects with Very-Large Eddy Simulation (VLES).

After the solving process is complete, post-processing extracts information out of the raw result data into a useful form. This can result in numerical values, such as total forces and moments, or visualizations, such as pressure/velocity maps, isosurfaces and streamlines.

3 State of the Art in Unsteady Aerodynamics

This chapter will summarize the current state of the art on unsteady vehicle drag. First, the observation of incident flow situation in real-world road and test track driving will be discussed. The next sections will explain how on-road unsteady flow can and has been recreated in simulation environments, both in the wind tunnel and in CFD. Finally, past findings concerning the effects of unsteady flow on drag compared to steady flow will be discussed.

3.1 Incident Flow Situation in the Real World

Saunders and Mansour [26] measured incident flow on different vehicles driving on two different test tracks, capturing flow angles and frequency spectra. Watkins and Saunders [27] used wind data to create a theoretical prediction of incident flow spectra on the road, which matched the aforementioned measurements. Cooper and Watkins [28] observed on-road turbulence without obstacles or traffic, considering different parameters such as natural wind and terrain. They applied then contemporary knowledge of general environmental turbulence to vehicle aerodynamics.

Wordley and Saunders [7] [29], Oettle et al. [30] and Schröck et al. [31] undertook road driving tests to investigate incident flow in real situations. These tests have provided results on incident flow angles, turbulence and frequency values for a number of different driving situations. The flow angle distribution is generally Gaussian, with the standard deviation being dependent on outside conditions such as presence of traffic or roadside obstacles. As for the turbulence data, an example can be seen in **Figure 3.1** for different driving cases. The majority of on-road flow situations fall between 2 and 6 % turbulence intensity and between 1 and 10 m in turbulent length scale. It is possible to further differentiate between specific cases. For example, during freeway driving with traffic turbulence intensity tends to be higher and length scale lower, while driving through city canyons has the opposite effect.

© The Author(s), under exclusive license to
Springer Fachmedien Wiesbaden GmbH, part of Springer Nature 2026
X. Fei, *The Impact of Unsteady Flow on Drag Measurements in Automotive Wind Tunnels*, Wissenschaftliche Reihe Fahrzeugtechnik Universität Stuttgart,
https://doi.org/10.1007/978-3-658-51766-3_3

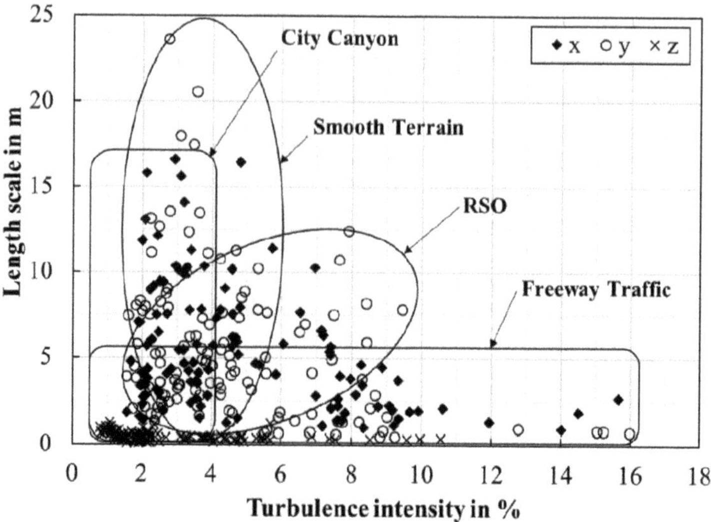

Figure 3.1:   Length scale and turbulence intensity for different driving cases [7]

Most commonly, multi-hole probes placed at the front or on top of the vehicle are used to capture incident flow data. Schröck et al. [32] used a specific arrangement of surface pressure taps as a flow sensor in addition, with the principle developed by Wagner called the vehicle-as-a-sensor method [33]. Furthermore, they separated their results by traffic density in addition to terrain, which was captured using an on-board camera. In addition to also concluding that on-road flow followed a Gaussian flow angle distribution (see **Figure 3.2**), Schröck et al. determined that incident flow frequencies above 2 Hz do not contain a large amount of energy and are therefore less relevant for vehicle dynamics.

Jessing [4] built upon the work by Schröck et al. by also undertaking highway measurements and complementing those with measurements on an empty test track with the test vehicle following another vehicle at different set distances. He confirmed the results of Schröck et al. in terms of incident flow angles, turbulence and frequency bands during his highway measurements. In addition, Jessing documented a reduced flow velocity in the direction of travel for the test vehicle due to traffic influence. **Figure 3.3** illustrates the reduction of flow velocity and the resulting possible rise in vehicle range at a travel speed of 100 km/h.

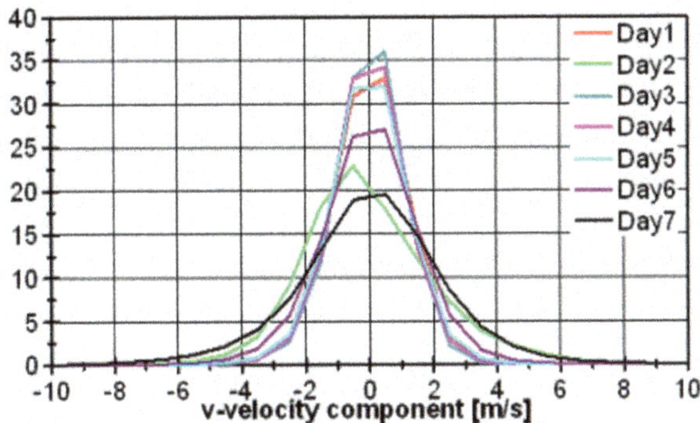

Figure 3.2: Measured Gaussian incident flow angle distribution, during highway driving tests, by Schröck et al. [32]

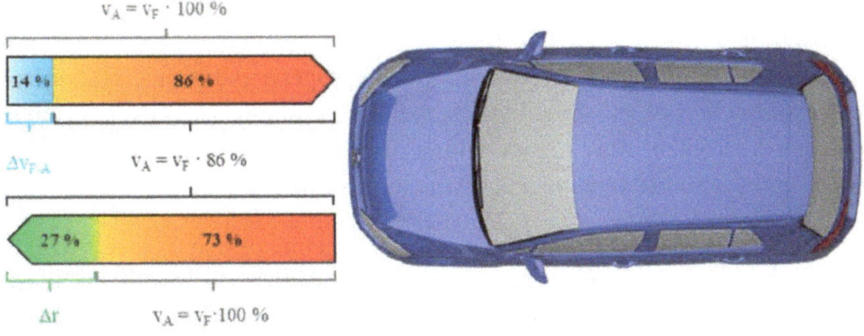

Figure 3.3: Average reduction of flow velocity compared to travel velocity of 14 % in traffic results in a 27 % increase in range compared to driving with no traffic [4]

Jessing's test track measurements also returned Gaussian flow angle distributions, with more frequent occurrences for higher flow angles the closer a test vehicle was following behind a leading vehicle. Furthermore, Jessing consolidated his measured turbulence data for both road and test track measurements. Most measurements fall into a range between 5 and 10 % turbulence intensity and between 0 and 2 m in turbulent length scale, which broadly agrees with Wordley and Saunders's results [7].

To summarize, the current knowledge on real-world flow is largely based on on-road or on-track measurements using cars equipped with flow probes. Flow velocity and direction, frequency spectra and turbulence values have been investigated. As a result, typical flow characteristics representative of different real-world cases, such as highway driving and city canyons, are known. In order to use this knowledge to advance aerodynamic design, however, these flow characteristics need to be replicated in simulation environments such as wind tunnels and CFD.

3.2   Unsteady Flow in the Wind Tunnel

The previous section has established typical incident flow parameters for on-road and on-track driving. The research has shown that realistic flow regimes for moving vehicles are very different from the steady-state, uniform flow that is commonly used in today's aerodynamic research and development. Therefore, the obtained flow parameters need to be replicated in a consistent and reproducible simulation environment to make proper aerodynamic evaluations under unsteady flow possible. Ideally, the test environment needs to create or simulate a turbulent flow regime with a turbulence intensity of up to ~10 % and length scales between ~1 and ~5 m (see section 3.1).

In the automotive wind tunnel, the main aerodynamic design goals have traditionally been a uniform and steady flow profile at the nozzle exit and in the test section [11]. As of the writing of this dissertation, no automotive wind tunnel has been designed and/or constructed exclusively for unsteady flow investigations. Therefore, in order to introduce transient flow, additional measures are necessary. Those fall under three broad categories:

1.   Stochastic calculations / Wind Averaged Drag
2.   Static turbulence generators
3.   Dynamic turbulence generators

3.2.1   Stochastic Calculations / Wind Averaged Drag

Stochastic methods initially obtain measurements in uniform and steady flow, but try to approximate real-world conditions by taking these measurements at different conditions and calculating a weighted average. Commonly, the result is known as "Wind Averaged Drag".

$$c_{D,unsteady} = \sum_{i=1}^{n} p(\beta_i) * c_{D,steady}(\beta_i) \qquad \text{Eq. 3.1}$$

with:

$c_{D,unsteady}$: stochastically calculated, wind averaged drag

$c_{D,steady}(\beta_i)$: steady-state drag at yaw angle β_i

$p(\beta_i)$: probability of occurrence for yaw angle β_i in observed unsteady flow

Buckley et al. [34] were some of the first researchers to apply stochastic calculations to vehicle aerodynamics. They used yaw angle distributions in wind data to calculate weighted average drag values from steady-state yaw measurements in order to analyze drag and fuel consumption of heavy trucks under real world flow. Since then, wind averaged drag, has appeared in a number of research undertakings in the commercial vehicle segment [35] [36] [37]. For passenger cars, Windsor discussed wind averaged drag in his work [38] and Howell et al. [39] suggested applying wind-averaged drag for more accurate vehicle certification during the WLTP driving cycle.

The application of wind averaged drag is not limited to natural wind. Since on-road and on-track measurements of flow angle distributions have become available (see section 3.1), it is equally possible to use those as weight factors instead. This is exactly what Jessing [4] did in his investigations, using the recorded flow angles from his on-road driving experiments to calculate averaged drag values.

Stoll [40] labeled his specific calculation method as "quasi-steady drag. Calculation of quasi-steady drag uses equation 3.1, with one-degree steps between discrete yaw measurements. In detail, the steps are as follows:

1. Determine the flow angle probability distribution of the signal in one-degree bands (e.g. from −0.5° to 0.5°, 0.5° to 1.5°, 1.5° to 2.5°, etc.)
2. Take drag measurements in a steady-state yaw sweep of the vehicle in one-degree steps (e.g. at 0°, ±1°, ±2°, etc.)
3. Calculate quasi-steady drag value by averaging the yaw sweep c_D values, weighted by the flow angle probabilities

While stochastic calculations are a simple and efficient way to incorporate real-world flow, they nonetheless do not take the transient part of the flow into account. Stoll [40], for instance, have shown that quasi-steady and transient measurements do not show the same results. Therefore, some way to create upstream turbulence in the wind tunnel is necessary.

3.2.2   Static Turbulence Generators

Static turbulence generators are static measures placed upstream from the model and the test section in a wind tunnel to create turbulent and unsteady flow.

Blockage bodies placed in or near the wind tunnel nozzle were used in the past to create upstream turbulence in the test section. These can be simple bodies or even complete vehicles. Watkins et al. [41] and Krampol et al. [42] used this method for aeroacoustics measurements (see **Figure 3.4**). However, this precludes measurements of aerodynamic forces due to the positioning of the blockage bodies.

Figure 3.4:   Upstream blockage vehicles in the test section. Left: [41]; Right: [42]

Grates are another simple way of introducing turbulence into the airflow, with the advantage of not interfering with force measurements. The resulting degree of turbulence can be controlled by changing grate parameters like mesh density or grid wire size. Watkins [43] was able to achieve turbulence intensities of $1.5 - 4\,\%$ streamwise and $2 - 5\,\%$ crosswise using different grates placed in the wind tunnel's last corner before the nozzle (**Figure 3.5**).

Figure 3.5:   Turbulence-creating grates in wind tunnel corner [43]

The Pininfarina wind tunnel employs a complex system of turbulence generators in its nozzle, designed by Cogotti [44], to simulate the natural wind in the
floor boundary layer. **Figure 3.6** shows this static, passive system, while Pininfarina's active system is described later in section 3.2.3. Schröck et al. [45]
used a similar, simpler system in the model scale wind tunnel, where they were
able to create a turbulence intensity of 3 % and length scales between
0.2 – 0.6 m.

Figure 3.6:   Turbulence generators in the Pininfarina wind tunnel [44]

Jessing [4] investigated shaped turbulence generators like Cogotti and Schröck
et al., as well as simple grating at the nozzle exit plane like Watkins (**Figure
3.7**). He achieved turbulence intensities of up to 6.5 % crosswise. However,
the achieved turbulent length scales (up to ~0.5 m) were too low to be comparable to his on-road and on-track flow measurements (length scales of ~1–
2.2 m).

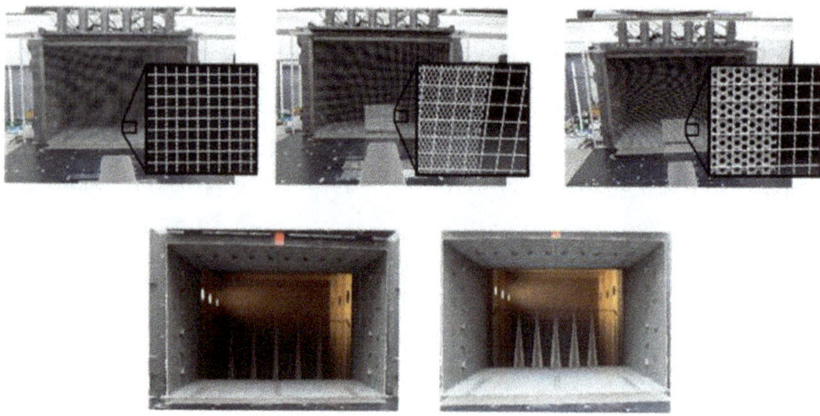

Figure 3.7:   Jessing's [4] static turbulence generators in the IFS model
scale wind tunnel. Top row: Grating at the nozzle exit with different mesh sizes and shapes at the nozzle exit. Bottom row:
Shaped turbulence generators inside the nozzle

The research using static turbulence generators has shown that while it is possible to create turbulence intensities comparable to on-road flow, it is impossible to achieve both realistic turbulence intensities and length scales at the
same time. This means that during wind tunnel measurements, either the average crosswise flow component or the large eddy sizes experienced by the
model will match the on-road flow, but not both. Additionally, it is impossible
to change the flow characteristics without modifying the static turbulence generators, increasing the time requirement for experiments. All in all, while static
turbulence generators offer an improvement in realism over stochastic calculations, they are still insufficient for investigations desiring comparability with
realistic flow.

3.2.3   Active Turbulence Generators

The previously mentioned unsatisfactory characteristics of static turbulence
generators has led to the implementation of active turbulence generators,
which are actively controllable and capable of generating predetermined unsteady flow signals.

Dominy and Ryan [46] [47] introduced a double jet system (**Figure 3.8**) in order to study wind gusts in the wind tunnel. A "main tunnel" provides regular, longitudinal flow while a "cross-wind tunnel" enters the test section through slots on one of the side walls. The crosswind incident angle is 30° and the duration of the gust can be controlled. The main limitation of this system are its fixed incident flow angle and its inability to create longer-running, high frequency signals.

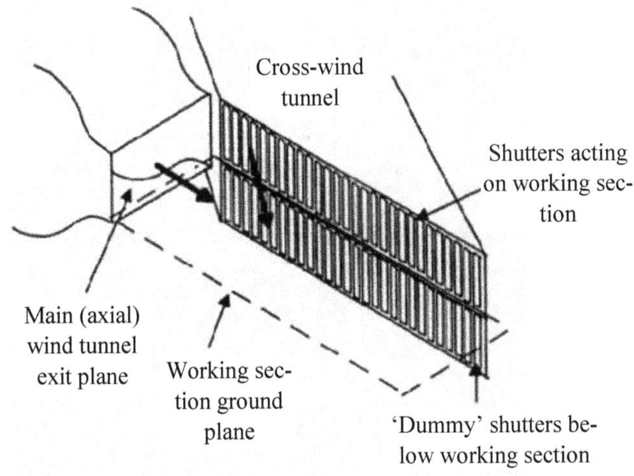

Figure 3.8:   Wind tunnel configuration in Dominy and Ryan's double jet system [47]

Mullarkey [48] and Passmore et al. [49] proceeded to develop a dynamic turbulence generation system for Loughborough University capable of creating different, customized unsteady flow signals. The system, as seen in **Figure 3.9**, consists of vertical, rotating airfoils in the test section, placed upstream from the model. They were able to achieve maximum flap angles of ±10° and a maximum frequency of 18 Hz.

Airfoils rotating along vertical axes have subsequently been the concept of choice in multiple wind tunnels. Among those are the Durham University model scale wind tunnel, the Toyota full scale wind tunnel and the University of Stuttgart model scale and full-scale wind tunnels.

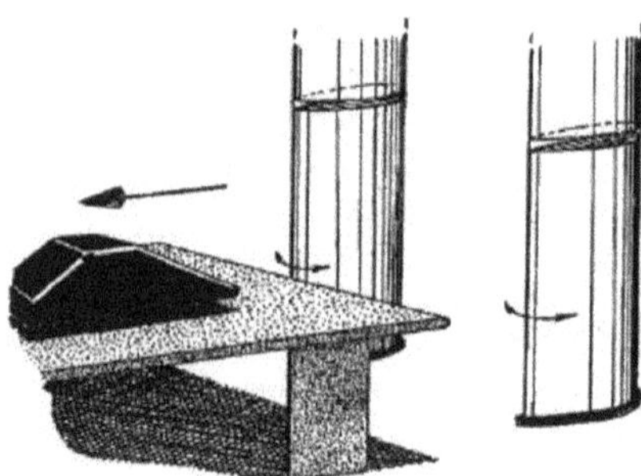

Figure 3.9:   Design sketch for Loughborough University's system [49]

The Durham University model scale wind tunnel employs two 600 mm chord airfoils following the side walls of the nozzle. Additionally, they use a unique system of variable shutters placed at the sides of the nozzle and the collector to reduce backflow in the plenum and therefore improve core jet flow quality (see **Figure 3.10** and **Figure 3.11**). The maximum achievable yaw angles and frequency were ±11° and 10 Hz respectively [50].

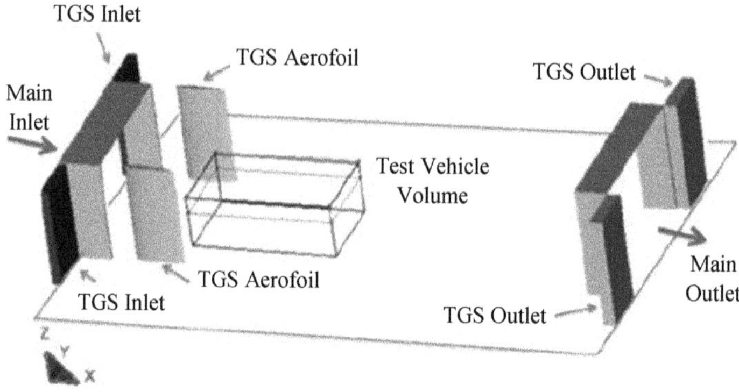

Figure 3.10:   CFD simulation model for the Durham University system [50]

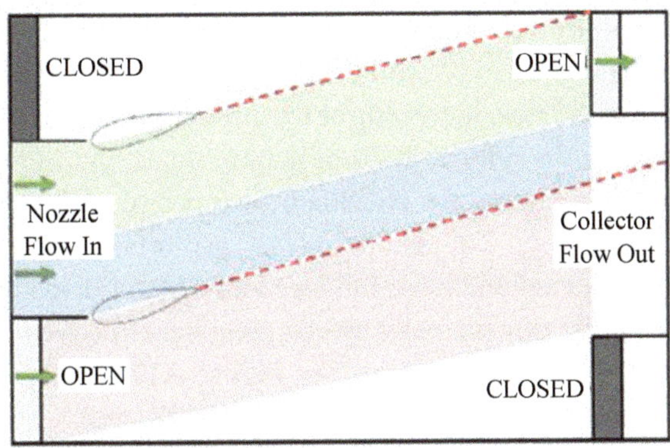

Figure 3.11: Principle sketch for the Durham University system [50]

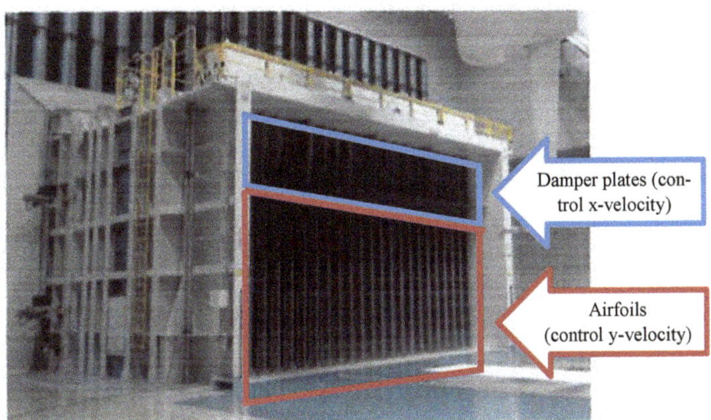

Figure 3.12: Toyota NWG with damper plates on top and vertical airfoils on bottom [51]

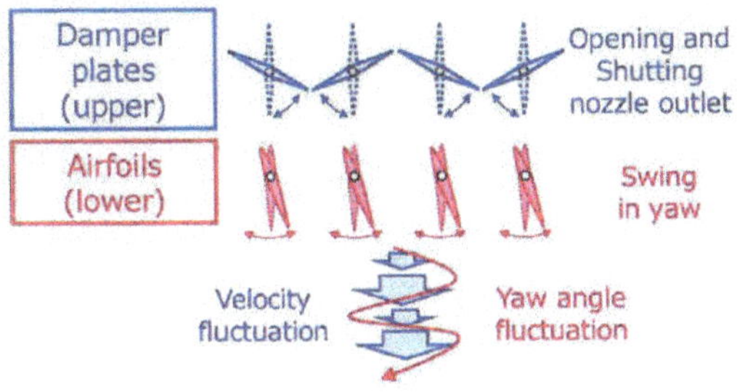

Figure 3.13: Principle sketch of Toyota NWG [51]

Instead of the two large chord airfoils in Durham's system, the Natural Wind Generator (NWG) in the Toyota wind tunnel has 24 smaller chord vertical airfoils placed across the nozzle to control the flow's yaw angle. These airfoils, however, only cover the bottom 70 % of the nozzle height. In the upper 30 % of the nozzle height, there are so-called "damper plates" that control the streamwise flow velocity. The damper plates also rotate along vertical axes, but are rhomboid-shaped and have full freedom of rotation as opposed to the airfoils. They alter the flow velocity by changing the effective nozzle exit area depending on their rotational position (see **Figure 3.12** and **Figure 3.13**). The NWG is thus unique in the sense that it can create discrete unsteady flow signals in yaw angle and flow velocity independently from each other. The maximum achievable flap angle is ±15°, streamwise velocity can be adjusted down to a minimum of 28% of original freestream velocity, and for both components the maximum frequency is 5 Hz [51].

The University of Stuttgart's system is called "FKFS swing" and was initially developed by Schröck [52]. The first system, seen in **Figure 3.14**, consisted of four airfoils at the nozzle exit that did not cover the whole nozzle height. Subsequently, FKFS swing underwent further development, and currently consists of 6 airfoils in the model scale [53] and 8 airfoils in the full-scale wind tunnel [12]. Yaw angle and frequency are limited to ±10° and 10 Hz for the full-scale wind tunnel [12], and ±10° and 12 Hz [53] for the model scale wind tunnel. A more detailed description of FKFS swing can be found in section 4.1.1.

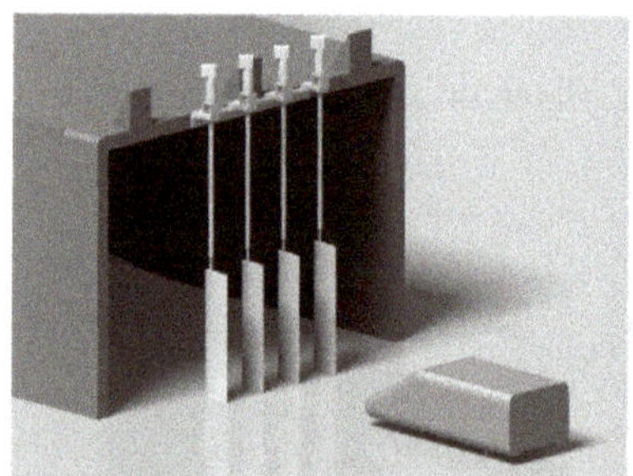

Figure 3.14:   Left: First iteration of FKFS swing by Schröck [52]; Right: Current FKFS swing system in the full-scale wind tunnel (only 7 out of 8 airfoils installed) [54]

Instead of rotating airfoils, Cogotti [55] uses two shaped flaps on each rotating axis to create the unsteady flow in the Pininfarina Wind Tunnel. The Active Turbulence Generation System (TGS) has three modes of operation: Static mode, with the flaps on each axis open at different angles from 15° to 49°; constant frequency mode, with the flaps moving in- or out of phase at a set frequency up to 1 Hz; and pseudo-random mode, where the flapping frequency is continuously varied. These modes enable the wind tunnel to simulate a multitude of different unsteady flow situations.

Figure 3.15: Pininfarina Active Turbulence Generation System (TGS), shown in [56]

It is also possible to create unsteady flow by using fixed airfoils with a movable trailing edge, like the German Aerospace Center (DLR) and Duke University show. Both systems are part of closed test sections in aerospace model scale wind tunnels. The DLR system [57] uses movable flaps that can reach maximum angles of ±20° at 50 Hz, while the Duke system [58] uses slotted cylinders to direct the airflow instead.

For all systems described in this section, it is important to note that the various maximum flap angles are the geometric angles between the flap chords and the longitudinal direction. Early research, such as [49] and [50], use "yaw angle" and "flap angle" interchangeably. However, Stoll [40] has shown that for most excitation frequencies the flap angle does not equal the true yaw angle of the flow in the test section. The exact relationship between flap and flow angle depends on the wind tunnel test section geometry and the individual turbulence generation system in question.

Active turbulence generation systems rectify the shortcomings of wind-averaged drag and static turbulence generators described in sections 3.2.1 and 3.2.2. With an active turbulence generator, unsteady flow signals with turbulence intensities and length scales consistent with real-world measurements can be created and exactly reproduced. However, introducing active turbulence generators often means significant cost both during installation and also during operation due to more complex flow situations and longer measurement times.

3.3   Unsteady Flow in Numerical Simulations

Numerical fluid simulations (CFD) of vehicles are commonly held in a practically blockage-free box environment, meant to simulate on-road flow without any obstructions. Compared to the wind tunnel, it is relatively simple to introduce unsteady flow into the environment – one needs to simply alter the inlet flow conditions to a variable, time-resolved one in the simulation setup. The important difference between CFD and wind tunnel testing lies in the simulation time, and by extension, the possible flow signals. It is not possible to simulate physical time scales that are similar to measurement times of typically 20 seconds or more, due to computational resource constraints. However, the physical time does need to be larger than for steady state simulations because it is necessary to capture a workable amount of the unsteady flow signal. Typically, steady state passenger car simulations need less than 1 second of physical simulation time to reach a state where evaluation is possible. For unsteady flow, depending on the exact method of turbulence creation, it can take from 2 seconds [59] to more than 7 seconds [60] of physical time. One way to reduce the simulation time is using a symmetry boundary condition at the vehicle centerline plane. However, this method works well only for highly symmetric vehicles like EVs and precludes the analysis of lateral wake oscillations [60].

While continuity solvers based on Navier-Stokes equations are able to solve unsteady flow, inherently unsteady solvers generally are better equipped for it. All past researchers present-ed in this section have used Simulia Power-FLOW for their investigations, which employs the Lattice-Boltzmann method. In addition, PowerFLOW offers an integrated turbulence generator that automatically creates a randomized transient flow at the simulation inlet ac-cording to prescribed turbulence intensity and length scale values [61].

Table 3.1:   Recommended turbulence values according to PowerFLOW best practice [61]

Turbulence value	Natural wind only	Natural wind and traffic
Tu_x	5.0 %	10%
Tu_y	5.0 %	10 %
Tu_z	0.75 %	1.5 %
TLS_x	2.5 m	2.5 m
TLS_y	2.5 m	2.5 m
TLS_z	2 m	2 m

Duncan et al. [62] and Gargoloff et al. [63] use the PowerFLOW turbulence generators for their investigations into the impact of turbulent flow on drag. PowerFLOW recommends typical on-road turbulence values (**Table 3.1**) that can be pre-set for the simulations. D'Hooge et al. [60] and Gaylard et al. [59], on the other hand, use unsteady signals they crafted themselves, such as step signals with set peaks or linear/triangular signals.

Commonly, the cell setup around the vehicle is not changed compared to steady-state simulations [59] [60] [62] [63]. Stoll [64], however, altered two regions near the vehicle and the vehicle wake, widening them out towards the back by 10° (see **Figure 3.16**). This way, he was better able to capture transient effects at the vehicle's sides, since his unsteady flow signals entail components with incident yaw angles up to 10°. Jessing [4] also follows this approach in his dissertation.

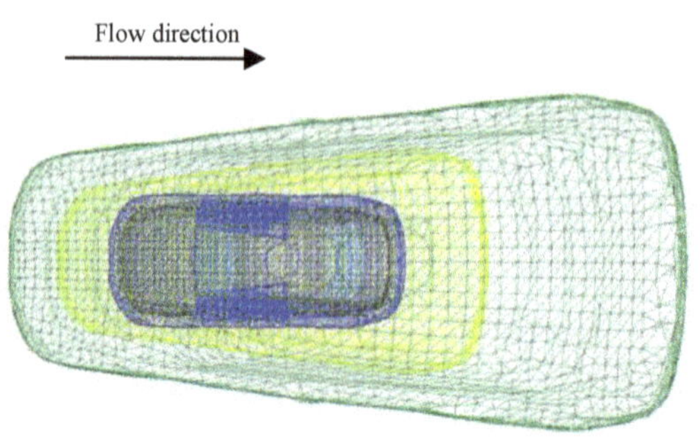

Figure 3.16:   Diverging variable resolution regions near the vehicle (1:4 scale DrivAer model), as described in [64]

In order to validate CFD setups, the results are usually compared with wind tunnel measurements results, as in D'Hooge's [60] work, for example. However, the setup in the wind tunnel is not the same as in CFD. For instance, the flow in the wind tunnel is not blockage-free and some interference effects are unavoidable. Therefore, Stoll [64] used a virtual model scale wind tunnel in his research (see **Figure 3.17**), based on the work of Fischer [65] [66] [67]. There, the unsteady flow is created by simulating the actual dynamic turbulence generators present in the real model scale wind tunnel. While this drastically increases simulation time and complexity, it offers the advantage of simulating the interaction of wind tunnel interference effects and unsteady incident flow.

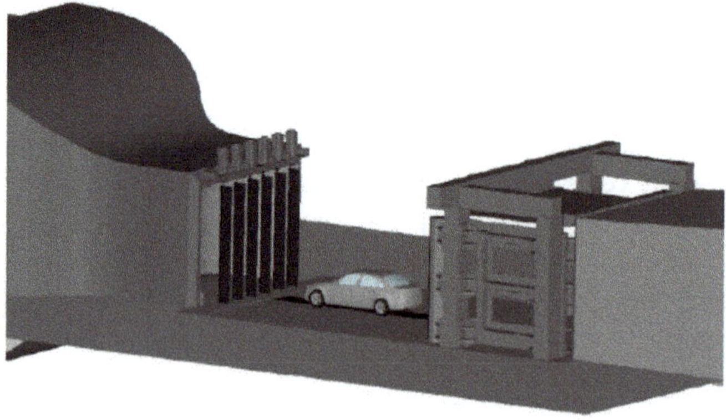

Figure 3.17:   Virtual model scale wind tunnel with dynamic turbulence generators [64]

Numerical simulations in a large box bypass a number of issues present in wind tunnel testing, such as interference effects and the necessity of dynamic turbulence generators. However, unsteady simulations are very time-consuming and costly and are also difficult to validate against measurements. They need to be used in conjunction with the wind tunnel for further research on unsteady flow.

3.4 Effects of Unsteady Incident Flow on Aerodynamic Drag

The effects of unsteady flow on vehicle aerodynamics have been investigated using the turbulence generation methods outlined in sections 3.2 and 3.3. Initially, research focused on areas such as flow field shapes, dynamic responses and side force coefficients. Watkins [43], Passmore et al. [49] and Schröck et al. [45] are some of the examples. Passmore was able to show clear differences in vehicle surface pressures between unsteady and steady state flow. His results are plotted in **Figure 3.18** and show much higher negative pressures on the side of the simplified vehicle model when transient flow is present.

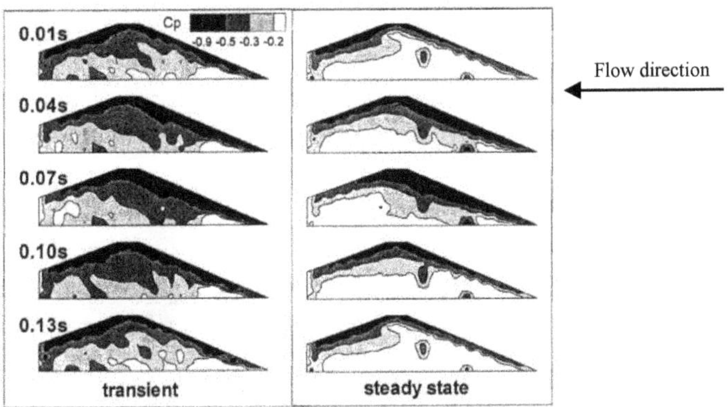

Figure 3.18: Surface pressure probe results on the side of a simplified reference body (Davis model, 1:6 scale) under transient and steady state flow by Passmore et al. [49]. Clear differences between pressures under transient flow (left) and steady-state flow (right)

More recent research, however, has started to focus on the effect unsteady incident flow has on drag, because out of all aerodynamic qualities it is drag that directly determines vehicle efficiency and especially an electric vehicle's range. Compared to drag measured under steady flow, unsteady drag is generally higher for most if not all vehicle geometries. This is evident even when using stochastic methods like in Windsor's measurements where he calculates wind-averaged drag [38] (see section 3.2.1). He measured 51 different cars of all body shapes in the wind tunnel. The wind-averaged drag results were, on average, higher than steady drag by 0.007–0.010 with a maximum delta of 0.015.

Cogotti [55] was one of the first to obtain drag results with a dynamic turbulence generation system, in the Pininfarina wind tunnel (see section 3.2.3). The Pininfarina system is able to operate both in static mode, where it acts as a static turbulence generator similar to meshes or other obstacles, and in dynamic mode. In both cases, vehicle drag was increased compared to steady, non-turbulent flow cases, with deltas as high as 0.025.

Numerical simulation results also agree that unsteady flow increases drag compared to steady flow. Gaylard et al. [59] and D'Hooge et al. [60] came to this conclusion after using their own created inlet flow signals, while Duncan et al. [62] and Gargoloff et al. [63] used the built-in PowerFLOW turbulence generator.

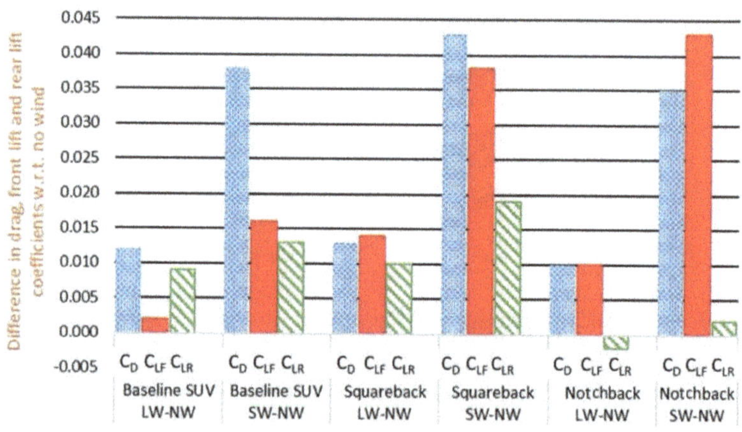

Figure 3.19:   CFD results from Gargoloff et al. [63], showing difference between unsteady (LW, SW) and steady (NW) flow

The drag results from Gargoloff et al.'s investigations are shown in **Figure 3.19** in the form of deltas between transient and steady state simulations. Here, "LW" and "SW" denote signals representing "Light Wind" and "Strong Wind" respectively, both unsteady flow signals. "NW" denotes "No Wind", meaning steady state results. The bars show the drag increase from steady to unsteady flow. The drag deltas (blue bars) are always positive, showing a bigger drag under unsteady flow, with the highest drag increase under unsteady flow at 0.043.

Recent research also compares unsteady drag obtained using stochastic methods (see section 3.2.1, i.e. quasi-steady drag) with unsteady drag obtained using active turbulence generators. Stoll [64] measured the DrivAer model in the model scale wind tunnel, with an active turbulence generation system applying unsteady flow signals representing light and strong wind respectively. The drag results (**Figure 3.20**), were consistently higher than those calculated from the same incident flow signals using quasi-steady drag. Therefore, Stoll concluded that real-world drag is underestimated if only steady yaw is taken into account and that the transience of the flow must be also considered. Yamashita et al. [51] obtained similar results in Toyota's wind tunnel using their Natural Wind Generator to produce the transient flow.

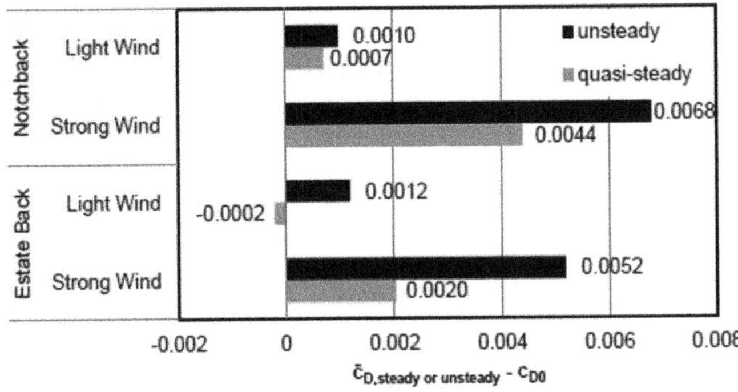

Figure 3.20:   Stoll's scale wind tunnel results [64] comparing unsteady drag versus quasi-steady (weighted average) results

In addition to the drag increase under unsteady flow, Gaylard et al. [59] have discovered that in some cases, the increased turbulence lessened or even entirely eliminated the beneficial effects of aerodynamic measures like wheel spoilers. The unsteady flow changed local wake shapes and therefore changed the effect of the spoiler on drag. Since the spoiler was initially developed in a steady flow environment, it was not optimized for unsteady flow and therefore lost some of its positive effects. Jessing [4] proceeded to thoroughly analyze the effect of unsteady flow on the efficacy of aerodynamic measures. He discovered that while some measures caused similar changes in aerodynamic forces in both steady and unsteady flow, other measures exhibited significantly different responses. One part that significantly changed its influence on drag was the separation edge of a hatchback's C-pillar. In steady state flow, it increased drag by ~0.008, but had no effect under unsteady flow.

To summarize, unsteady flow environments generally increase aerodynamic drag, sometimes significantly. It is necessary to undertake research using active turbulence generation, either in the wind tunnel or in CFD, because quasi-steady investigations using weighted steady yaw measurements do not consider transient effects. Furthermore, aerodynamic measures like spoilers and separation edges can have different effects on drag depending on whether they are measured in unsteady flow.

What is not known yet, however, is the following:

- How different vehicle geometries and different unsteady flow environments interact in their influence on drag
- How changes in the simulation environment (wind tunnel geometry or CFD setup) influence resultant unsteady drag

These points are the main focus of the work presented in this thesis. The following chapters will expand on those points using experimental and numerical investigations.

4 Experimental Investigations and Results

The experimental investigations of this work were undertaken in the University of Stuttgart Model Scale Wind Tunnel. First, the experimental setup will be described. Thereafter, measurement results will be presented, covering pressure gradients, interference corrections, absolute drag results and quantification of transient flow influence on drag.

4.1 Experimental Setup

4.1.1 Wind Tunnel and Turbulence Generation System

All experimental investigations were carried out in the University of Stuttgart's Model Scale Wind Tunnel, operated by FKFS. Henceforth, it will be referred to with its abbreviation MWK. The wind tunnel has closed air circuit with a 350 kW fan, enabling wind speeds of up to 80 m/s (288 km/h). The nozzle cross-section area is 1.7 m^2, opening up into a ¾ open test section with a length of 2.6 m. A state-of-the art ground simulation system, consisting of a 5-belt system and various boundary layer control systems, are installed in the floor of the wind tunnel, as is a rotating turntable to vary incident yaw angles. The dynamic underfloor balance can measure both steady and unsteady aerodynamic forces and moments [53]. All measurements were carried out at a flow velocity of 45 m/s (162 km/h). The aforementioned ground simulation systems were inactive for all measurements, in order to focus on the effects of the unsteady incident flow.

A 6-component underfloor balance measures forces and moments in a vehicle-fixed coordinate system. Drag is always measured parallel to the vehicle heading, side force is measured perpendicular to the vehicle heading, and positive yaw angles occur when the balance is turned clockwise (from the top view). The latter is because the yaw angle's sign depends on the direction of the incident flow relative to the vehicle heading. This direction is opposite to the turntable rotating direction, thus the opposite signs. **Figure 4.1** shows this visually.

© The Author(s), under exclusive license to
Springer Fachmedien Wiesbaden GmbH, part of Springer Nature 2026
X. Fei, *The Impact of Unsteady Flow on Drag Measurements in Automotive Wind Tunnels*, Wissenschaftliche Reihe Fahrzeugtechnik Universität Stuttgart,
https://doi.org/10.1007/978-3-658-51766-3_4

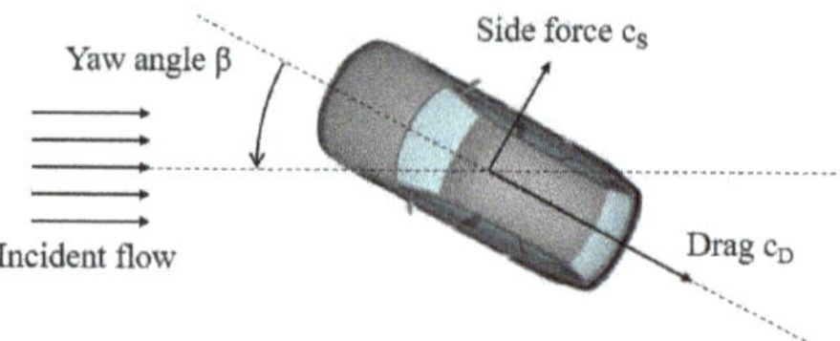

Figure 4.1:   MWK measurement directions, visualized

The FKFS swing system installed in the MWK is an active turbulence generation system that was briefly touched upon in section 3.2.3. It consists of six identical, vertical wings, placed at the nozzle exit (**Figure 4.2**). It can operate at a maximum wind speed of 50 m/s with a maximum yaw angle and frequency of 10° and 12 Hz respectively [31] [68]. During actuation, all wings move synchronously, meaning the flap angle of all wings are identical at the same point in time.

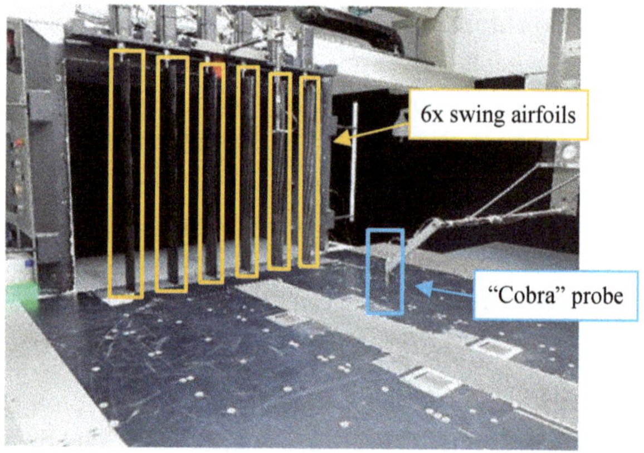

Figure 4.2:   MWK test section with FKFS swing and "Cobra" transient flow measurement probe installed [69]

While unsteady measurements were taken with the model pointing in the flow direction and the sideward flow component provided by FKFS swing, the steady-state yaw sweeps were taken with a 0° flap angle. In that case, the turntable was used to provide a steady yaw angle.

As outlined in section 2.4.2, interference corrections for wind tunnel measurements are generally necessary to achieve results comparable to the on-road case. The MWK can apply the current state-of-the-art method, the Two-Measurement correction [16]. In order to do so, an alternative test section setup is required to create a second static pressure gradient. This is achieved by mounting stagnation bodies to the walls of the collector, blocking the flow and thereby raising the static pressure in the rear of the test section.

There are two size options for the stagnation bodies, extending into the flow path by 48 mm ("S") and 76 mm ("L") respectively. The larger body creates a bigger pressure gradient close to the collector. As a result, three different static pressure distributions are possible in the test section, enabling correction validations as described in section 2.4.2. **Figure 4.3** shows a sketch of the stagnation body geometry, while **Figure 4.4** shows the resulting static pressure distributions.

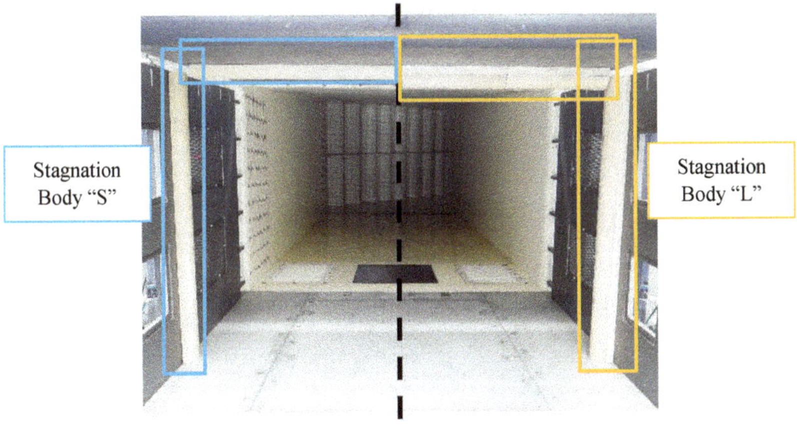

Figure 4.3: Stagnation bodies in MWK collector. Left half shows the 48 mm body (Size "S"), right half shows the 76 mm body (Size "L")

Measurements in the wind tunnel are split into empty test section measurements and vehicle measurements. Empty test section measurements are necessary to capture the test section flow conditions, as well as record the static pressure gradients necessary for interference corrections. A Prandtl probe is used to measure the static pressure distribution in 100 mm steps along the wind axis, taking 30 second averages per measurement point. Afterwards, the front and rear parts of the results are smoothed out using a fifth-degree polynomial. For the flow direction, a fast-response cobra probe (see also: section 2.4.1, pictured also in **Figure 4.2**) was placed in the middle of the turntable (x = 0, y = 0) at z = 250 mm above ground to measure the transient flow velocity components u, v, w, for 60 seconds per flow signal, at a resolution of 1250 Hz. Both these measurements were taken for the base, steady flow case as well as every individual unsteady flow signal that will be described in section 4.1.2.

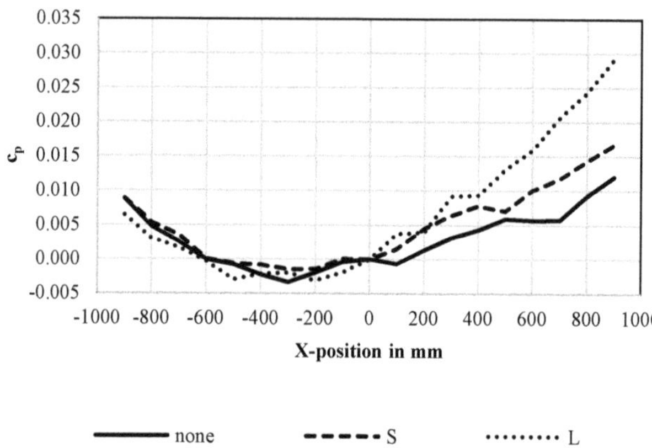

Figure 4.4:   Exemplary static pressure distributions in the MWK for one incident flow signal but different stagnation bodies ("none": no stagnation body; "S": 48 mm size; "L": 76 mm size)

Vehicle measurements capture the drag responses of the different models to different incident flow signals. Measurements for steady flow situations took place over 10 seconds, while those for unsteady flow situations took place over 30 seconds. The resultant force and moment coefficients were then averaged to obtain one set of mean results for each data point. Measurements with select vehicle models and signals were also undertaken at double and triple measurement times (i.e. 60 seconds and 90 seconds for unsteady measurements) and resulted in drag values within the usual measurement error of ±0.0005 compared to the originally selected measurement timeframes, thereby validating their selection.

4.1.2 Unsteady Incident Flow Signals

The unsteady flow signals applied during the experiments belong to two distinct types, namely filtered random noise signals (henceforth: noise signals) and sine signals. The noise signals are generated using a random number distribution for the flap angle over time, cutting off angles above 10° for operational safety, and then applying a first-order Butterworth low-pass filter with the cutoff frequency f_c. The resulting flow angle distribution (**Figure 4.5**) shows an approximately normal (Gaussian) curve. As such, a noise signal is characterized sufficiently using its standard deviation σ to quantify its deviation from the freestream direction, in addition to f_c.

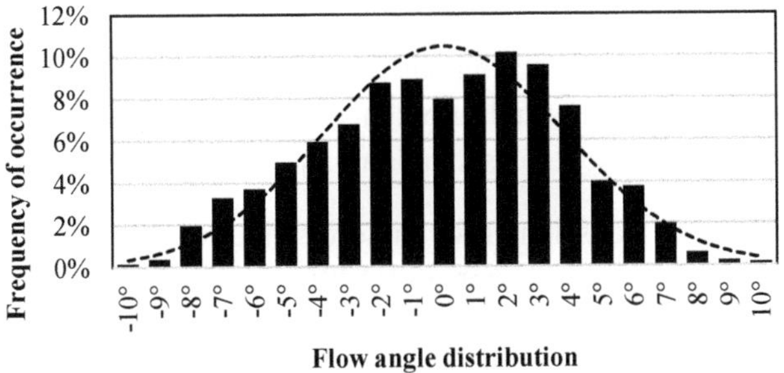

Figure 4.5: Flow angle distribution (measured in empty test section) for a filtered random noise signal (f_c = 7.5 Hz; σ = 3.8°). The dotted line represents an ideal Gaussian distribution

The sine signal (**Figure 4.6**), on the other hand, is a simple, one-frequency signal characterized by its frequency f and its sine amplitude A. In addition, it is also possible to determine the standard deviation σ for the sine signal using the formula

$$\sigma = A/\sqrt{2} \qquad\qquad \text{Eq. 4.1}$$

Its flow angle distribution is very different from the noise signals, with the angles close to its amplitude being the most and the angles around 0° being the least common.

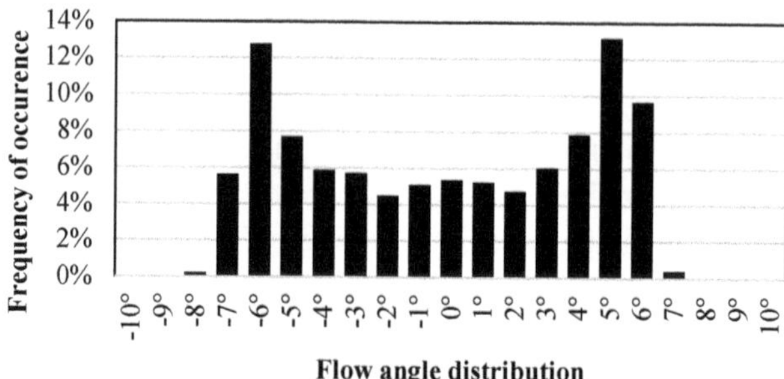

Figure 4.6:   Flow angle distribution (measured in empty test section) for a sine signal (f = 7.5 Hz; A = 6.3°)

Appendix A1 shows the characteristic data of all flow signals created, including their Strouhal numbers and turbulence values. In this work, 16 noise signals and 20 sine signals were used, adding up to 36 distinct signals in total.

The noise signals' Gaussian flow angle distribution matches up with real-world incident flow measurements [32] [70] as described in section 3.1 (compare with **Figure 3.2**). Furthermore, their standard deviation turbulence numbers line up with recently observed on-road flow made by Jessing et al. [70] (see Appendix A1). Noise signals are therefore well suited to research and predict aerodynamic performance during realistic use cases.

Since the sine signals' flow angle distributions are not Gaussian and differ substantially from real-world measurements, they cannot be used to accurately judge on-road aerodynamic performance. However, this is not their goal. The advantage sine signals hold over noise signals is their simplicity. They have one single frequency, compared to the many frequencies of a noise signal. Furthermore, a sine signal has a short (under 1 s), constant period after which the signal repeats. All in all, sine signals are useful for investigations into fundamental phenomena because of consistency and low variability.

4.1.3   Vehicle Models

The vehicle models chosen for the measurements differ in various geometric traits, such as their overall size, level of detail and rear end configurations.

This allows conclusions to be made regarding geometric influences on unsteady drag. Furthermore, the models are all open-source reference bodies, allowing others to continue and/or repeat the present work.

The chosen models are as follows:

1. SAE Reference Body (abbrev. "SAE Body") [71]
2. SAE Reference Body, Type K (abbrev. "SAE Type K") [72]
3. DrivAer [22] [73]
4. AeroSUV [23]

All models are scaled 1:4, which avoids recalculation of dimensionless values dependent on scale. They all have interchangeable rear ends that allow setup in either a notchback or a squareback configuration.

During the investigations, all measurements and simulations were performed with a closed grill, so there was no airflow through the engine compartments of the models.

The SAE Body (**Figure 4.7**) is a simplified model meant to represent a passenger car. Despite its lack of details, the SAE Body offers some typical automobile features important in aerodynamic research, such as distinct A-pillars, C-pillars (for the notchback/fastback rear end), sharp trailing edges at the rear, and an underbody diffuser [71].

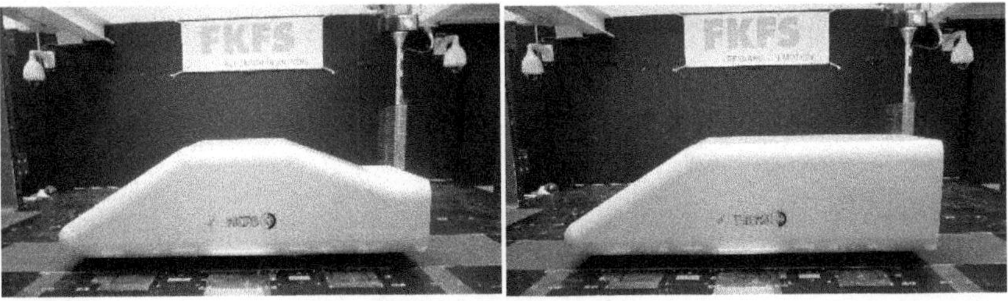

Figure 4.7:   SAE Body, notchback (left) and squareback (right)

Kuthada et al. [72] modified the basic design of the SAE Body in order to evaluate the effects of ground simulation systems on cooling air drag, creating the SAE Type K (**Figure 4.8**). The dimensions and the surface geometry remained the same, except for the newly created wheelhouses containing rotatable wheels. Furthermore, a simple engine compartment was included in the model [72], but is not used in this work. The Type K model shares its rear end interchangeability with the original SAE reference body.

Figure 4.8:   SAE Type K, squareback configuration [74]

The DrivAer model (**Figure 4.9**) was initially developed by the Technische Universität München in cooperation with Audi AG and BMW group and introduced in 2012. Its aim was to offer a realistic generic car model for aerodynamic investigations. The DrivAer is a blend of the then current Audi A4 and BMW 3 series models [22]. It possesses a high level of detail, as seen in the detailed underbody, mirrors, door handles, minute indentations in the grille, doors, and hood areas, and other features.

The DrivAer possesses an additional fastback rear end option compared to the SAE Body. During the present investigations, the fastback was omitted, since the notchback and fastback tend to behave fairly similarly. In order to enable investigations on cooling airflow, the open grille DrivAer was introduced, offering a realistic cooling airflow and engine [73]. The grilles can also be replaced with solid pieces to close off the internal flow.

Figure 4.9:   DrivAer, notchback (left) and squareback (right)

Compared to the SAE Body's highly simplified geometry, the DrivAer model possesses a high level of detail. In order to close the gap and to isolate the effects of a simplified and sharp-edged rear end geometry, new rear ends for the DrivAer have been created. These rear end geometries are based on the SAE Body and mimic their rear trailing edges. In the notchback model's case, the C-pillar geometry has also been imitated (see **Figure 4.10**).

Figure 4.10: DrivAer models with different, newly created rear end modifications [74] Top Left: Notchback modification (A); Top Right: Notchback Modification (B); Bottom: Squareback modification

While the DrivAer model has been readily adopted for aerodynamic research, a similarly detailed and realistic open-source model has not been available for the SUV category, which has been increasing in popularity. To rectify this, FKFS and Roechling Automotive have developed the AeroSUV model (**Figure 4.11**) [23]. Its dimensions were derived from typical contemporary mid-class SUVs, such as the Audi Q5 or BMW X3. The model's geometry was further refined in a way that it performed similarly to the aforementioned mid-class SUVs aerodynamically. The AeroSUV is based on the DrivAer geometry and shares many traits with it, such as internal structures to allow cooling airflow, a detailed underbody, mirrors and rotating wheels. It shares the same detachable rear ends with the DrivAer, meaning it can also be configured as a notchback, fastback or a squareback vehicle, with the squareback representing standard SUVs and the other configurations representing SUV coupes. As with the DrivAer, only the notchback and squareback configurations were investigated here.

Figure 4.11:   AeroSUV, squareback configuration in the MWK test section [23]

4.2   Static Pressure Gradients and Interference Corrections

After establishing the experimental setup, the following section will show the static pressure gradient results measured while under the aforementioned incident flow signals. Furthermore, this work represents the first instance of applied interference corrections in the context of unsteady incident flow.

4.2.1   Static Pressure Gradient Results

As described in section 2.4.2, every open-jet wind tunnel has a unique, non-constant static pressure distribution along its wind axis. Most commonly, a non-zero pressure gradient is caused by nozzle or collector interference [14] or boundary layer control mechanisms [75]. Insufficient adaptation of the static pressure in the nozzle to the ambient pressure in the plenum causes the nozzle gradient, while the jet's shear layer stagnating at the collector entry creates the collector gradient [14]. Both interference effects are dependent on the wind tunnel geometry. Boundary layer control systems, such as suction mechanisms, can change the static pressure immensely at the vehicle front [75]. However, since the experiments here were carried out without ground simulation, these effects have not been considered.

To quantify the effects of the flow signals on the static pressure gradient, it was measured for every flow signal as described in section 4.1.2. **Figure 4.12** shows an example result.

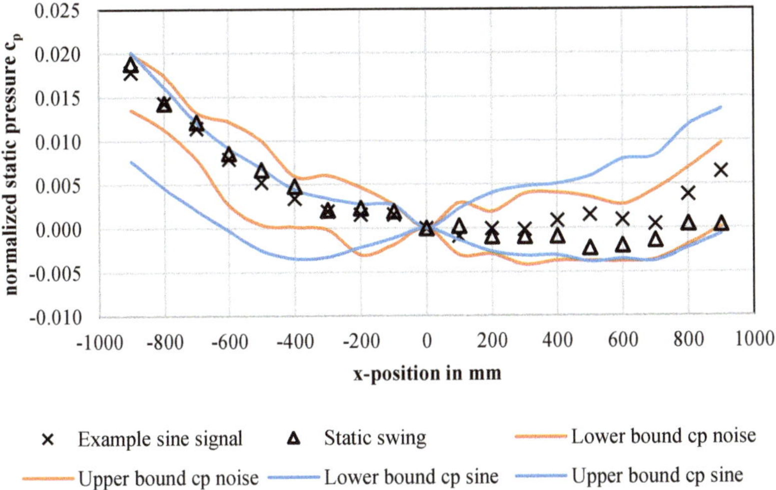

Figure 4.12:   Static pressure dist. in the empty test section, normalized to turntable center (sine signal; $f = 1.5$ Hz, $A = 4.4°$). Colored lines represent upper and lower bounds measured with unsteady signals to show gradient range

The empty test section pressure gradient influences the measured drag by superposing its pressure delta from the vehicle wake area to the vehicle front onto the vehicle surface pressures. This pressure delta is therefore a sufficient quantifier for the gradient influence. For the present investigations, the pressure delta was defined as:

$$\Delta c_p == c_p(x_2 = 775\ mm) - c_p(x_1 = -537.5\ mm) \qquad \text{Eq. 4.2}$$

To define Δc_p for a set of measurement, it is sufficient to choose x_1 and x_2 in such a way that they can represent the front end and the end of the near-field wake of a vehicle, respectively. The specific values here originate from the

SAE Body (see section 4.1.3). x_1 is the position of the model's front when placed in the test section, whereas x_2 is the position a quarter-vehicle length behind the SAE body. Wake calculations detailed in [15] and [16] show that this x_2 value reasonably represents the SAE Body's near wake end position.

For the pressure distribution shown in **Figure 4.12**, the pressure delta is $\Delta c_p = -3.8 * 10^3$, or -3.8 counts. The negative value shows a decreasing static pressure towards the collector, meaning that the pressure distribution will artificially add to the measured c_D values. For a positive delta value, the opposite would be true.

In order to observe the differences in pressure delta with different signal parameters, the deltas resulting from all sine signals (without stagnation bodies) are plotted in **Figure 4.13**.

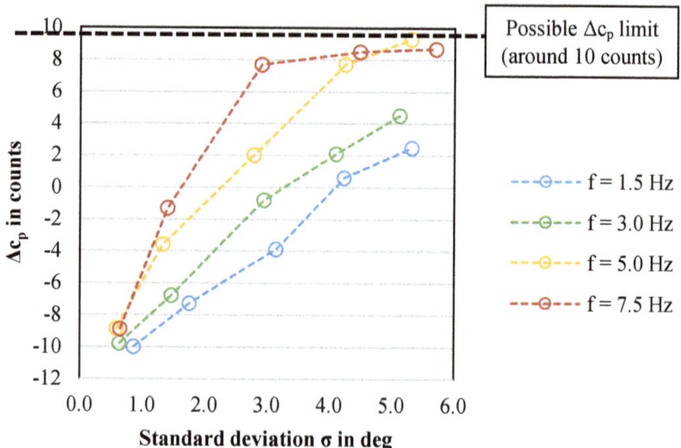

Figure 4.13:   Measured pressure deltas in the empty test section for all investigated sine signals, without stagnation bodies

For every frequency, the pressure delta increases with increasing standard deviation of the signal. Only at high σ for $f = 7.5$ Hz (red) there is an exception. From a standard deviation of about 3° upward, pressure delta barely increases with further increase in σ. Furthermore, the $f = 5$ Hz curve (yellow) also flattens out slightly past σ = 4°. These observations suggest that there is an upper limit for pressure deltas caused by sideward flow, in this case possibly around

10 counts. Increasing pressure deltas under unsteady flow are caused by increasing pressure near the collector, which in turn is caused by:

- Lower local flow velocity: High yaw angle unsteady flow distributes the wind tunnel air mass flow over a larger area at lower speeds
- Stronger stagnation of wind tunnel jet at collector entry (collector is designed for steady-state flow and too small for the larger area the air mass flow is spread over)

Neither of these phenomena can be amplified endlessly by an increasing flow angle, which explains the pressure delta plateau. At a sufficiently large σ, the airflow will:

- Strongly interact with the plenum walls, preventing the airflow from spreading over a larger area
- Completely enclose the collector walls, thereby reaching maximum stagnation at the collector entry

Next, the different frequency curves show that increasing signal frequency leads to increasing pressure deltas. This is more pronounced at higher signal standard deviation. The exception at high signal frequency and standard deviation can be explained by the pressure delta plateau discussed previously

After observing the sine signals in **Figure 4.13**, **Figure 4.14** shows the same pressure deltas resulting from the noise signals.

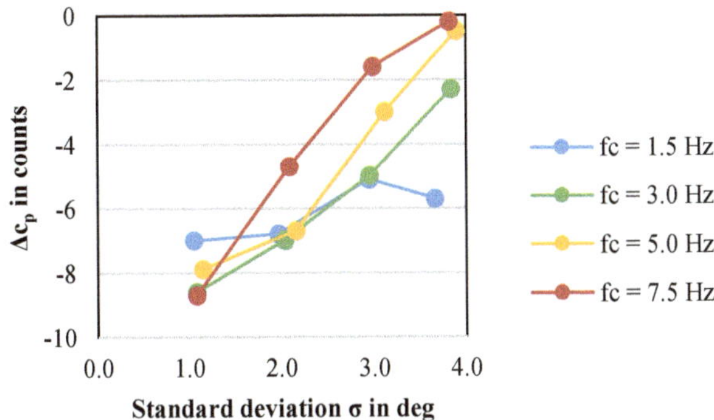

Figure 4.14: Measured pressure deltas in the empty test section for all investigated noise signals, without stagnation bodies

For the noise signals measured, there is no pressure delta plateau like for the sine signals. The noise signals investigated have lower σ than the sine signals and lower maximum Δc_p, because swing is unable to create higher σ values for noise signals before operational limits are reached. Both values are sufficiently far away from the rough limits observed for the sine signals previously.

The positive correlation between the characteristic frequency (here: cutoff frequency f_c) and the pressure delta also exists but is much weaker compared to the sine signals'. The pressure deltas of signals with the same flow angle but different f_c tend to be similar especially at low frequencies. Since a signal with the cutoff frequency f_c also contains all frequencies below f_c, the signals are more similar to each other than the sine signals are. This explains the weaker correlation compared to the sine signals.

All in all, it has been established that generally, pressure deltas between the front and back of the wind tunnel are correlated to unsteady signal type, frequency and standard deviation. Since the pressure delta directly changes measured, uncorrected drag, these observations support the application of interference corrections for unsteady drag. This will be discussed further in section 4.2.3, after a valid correction method is established in section 4.2.2.

4.2.2 Validity of Interference Corrections under Unsteady Flow

The previous section has established that the incident flow signal created by FKFS swing strongly influences the static pressure gradient in the test section, which needs to be accounted for using interference corrections. The Two-Measurement correction is the state-of-the-art method for correcting steady-state measurements (see section 2.4.2) and has been applied to unsteady measurements [64]. However, to date, there has been no validation or accuracy evaluation for this method in an unsteady environment. This section will therefore evaluate the Two-Measurement correction using MWK measurements and determine how to proceed in order to ensure correction accuracy.

The Two-Measurement correction needs two measurements in different wind tunnel configurations to iterate a corrected c_D value. However, in order to validate the correction method, additional wind tunnel configurations must be considered. With the two different stagnation bodies ("S" and "L", see section 4.1.1), along with the standard test section without stagnation bodies ("none"), three different wind tunnel configurations in the MWK can be represented. Therefore, three different correction options are possible, each using two of the three wind tunnel configurations:

- Correction using "none" and "S" ("none+S")
- Correction using "none" and "L" ("none+L")
- Correction using "S" and "L" ("S+L")

However, apart from the different wind tunnel configurations, Cooper [76] has established another requirement for error-free gradient corrections. The two wind tunnel configurations used for the correction need to have "self-similar" empty test section static pressure distributions. Two pressure distributions are considered self-similar if one of them can be mathematically transposed into the other by moving it along the wind axis, then multiplying it with a constant factor [76]. An in-depth explanation of self-similarity can be found in Appendix 0.

Applying the criterium of self-similarity eliminates the correction option "S+L" because it is comprised of two pressure distributions with artificially high gradients near the collector (created by stagnation bodies). The other two correction options only use one high gradient pressure distribution each. Pressure distributions with high positive gradients close to the collector increase the error caused by lacking self-similarity, as explained in detail in Appendix 0.

The availability of multiple different options for the correction allows for the validation of the correction method: If the correction method is working as intended, the multiple corrected drag values resulting from multiple options must be within a small margin. This margin (Δmax) is then used to quantify the accuracy of the correction method. A maximum Δmax of under 0.002 (2 drag counts) is generally viewed as good validity [17]. This technique has been utilized for correction validations in the past, such as during wind tunnel correlation measurements [19]. It can be used for both steady-state as well as unsteady measurements.

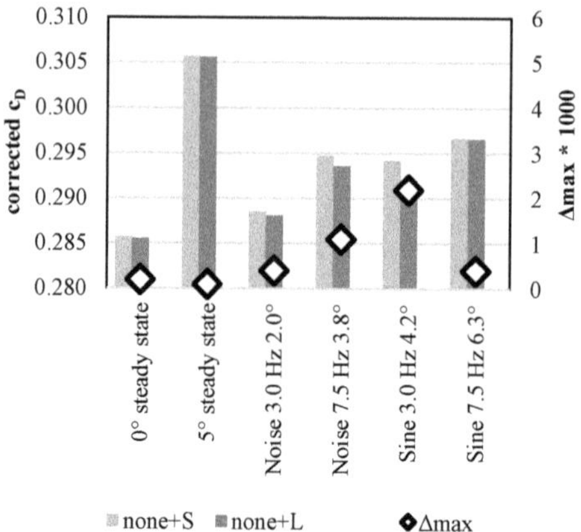

Figure 4.15:   Selected, corrected c_D results from the DrivAer notchback model, using the two correction options "none+S" and "none+L"

Figure 4.15 shows selected results from the drag measurements. The steady state results are very accurate, with the delta between the correction options less than half a drag count. Despite the pressure distributions not being self-similar, the unsteady corrections are still accurate, with only one measurement slightly above a delta of two counts. The higher error for unsteady corrections compared to the steady corrections results from the comparatively high pressure gradient between model and collector under unsteady flow. It compounds the error introduced by the lacking self-similarity of the gradients.

To prove that the elimination of the correction option "S+L" is the correct choice, the next figures will show the same c_D results and their correction margins Δmax involving "S+L". **Figure 4.16** shows that after adding the "S+L" option to **Figure 4.15** the deltas increase considerably, with all but one of the unsteady deltas higher than 2 drag counts.

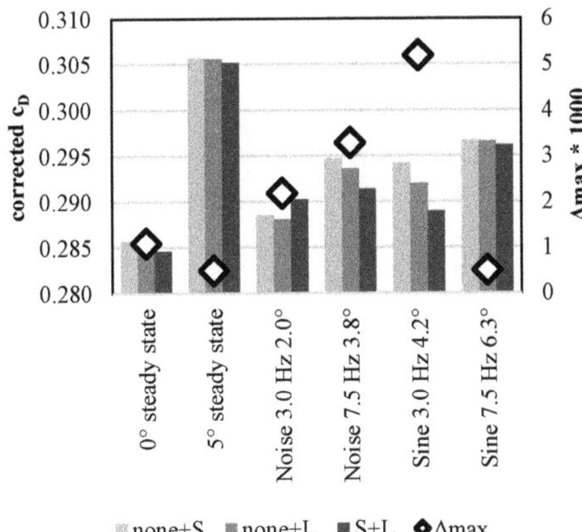

Figure 4.16: Selected, corrected c_D results from the DrivAer notchback model, with all three correction options (same raw data as **Figure 4.15**)

The maximum delta between three options (**Figure 4.16**) is, of course, always equal or higher than the maximum delta between two options. However, even when only considering two options, the combination shown in **Figure 4.15** provides the best accuracy. **Figure 4.17** and **Figure 4.18** show that whenever the correction option "S+L" is involved, the delta between the corrected results remains largely above 2 drag counts. Therefore, the results would not meet the required accuracy criterium.

To summarize, when applying interference corrections while using an active gust generation system like FKFS swing, it is important to consider the self-similarity of the resultant static pressure distributions. When using measures to artificially raise the pressure gradient near the collector, it is better to only moderately modify the gradient, because the unsteady flow will further increase the gradient and negatively influence self-similarity. For this work, all the c_D values presented from here on will be corrected for interference effects unless otherwise stated. The value will be the average of the correction results with "none+S" and "none+L", ensuring correction accuracy.

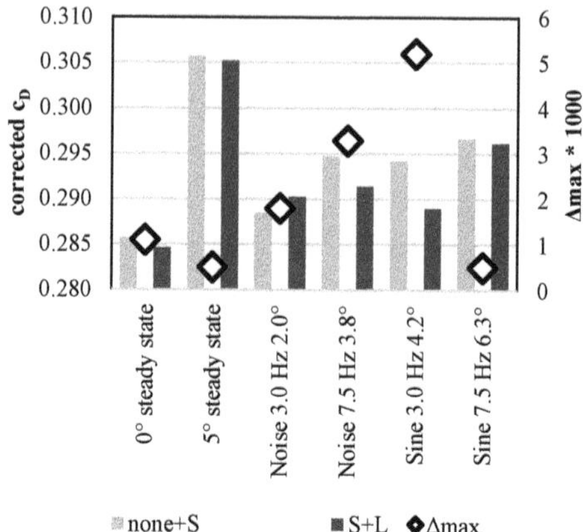

Figure 4.17:   Selected, corrected c_D results from the DrivAer notchback model results, using the correction options "none+S" and "S+L" (same raw data as **Figure 4.15**)

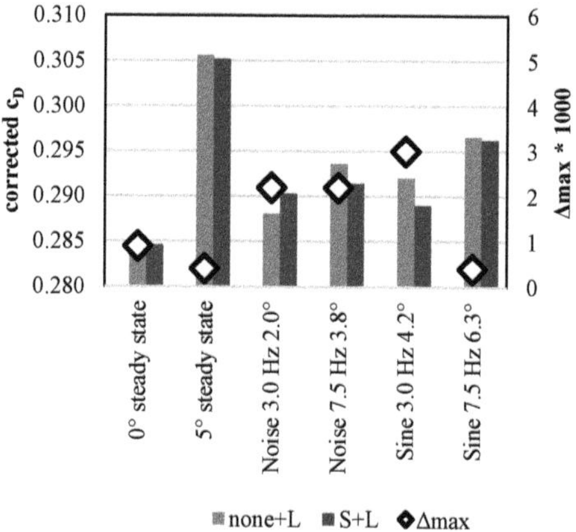

Figure 4.18:   Selected, corrected c_D results from the DrivAer notchback model results, using the correction options "none+L" and "S+L" (same raw data as **Figure 4.15**)

4.2.3   The Necessity of Interference Corrections

Having validated the Two-Measurement Correction method for unsteady flow, this section will underline the importance of interference correction in the context of unsteady investigations. For this, selected results from MWK measurements will be shown, corrected with the method established in the previous section 4.2.2.

Currently, corrections are not part of most standard automobile wind tunnel measurements, which is a valid approach as they only use steady-state flow configurations. Many modern wind tunnels would only need a steady-state drag correction in the magnitude of a few drag counts. To illustrate this, **Figure 4.19** shows steady-state yaw sweep results of the DrivAer notchback model in the MWK. While there is a significant difference between corrected and uncorrected c_D, the difference between the two remains virtually identical for all yaw angles shown. This shows that while the absolute drag value may not match up with blockage-free flow, the differences between configurations typically remain the same and can be analyzed without correcting for wind tunnel interference.

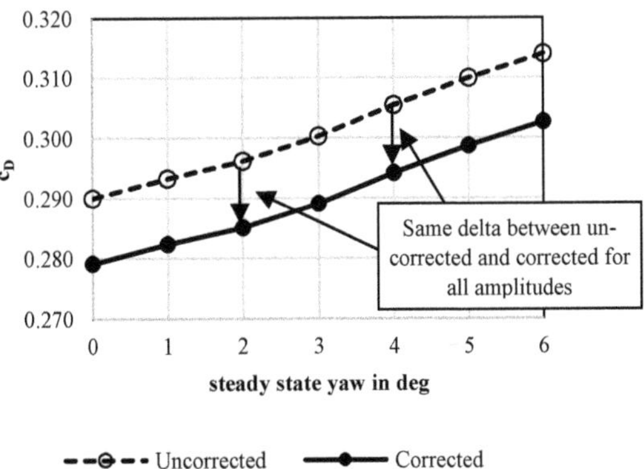

Figure 4.19:   DrivAer notchback, steady-state yaw sweep, uncorrected vs corrected drag

The same method used in **Figure 4.19** was also applied to unsteady measurements, plotting uncorrected and corrected drag over signal standard deviation. Results for both noise and sine signals can be seen in **Figure 4.20**.

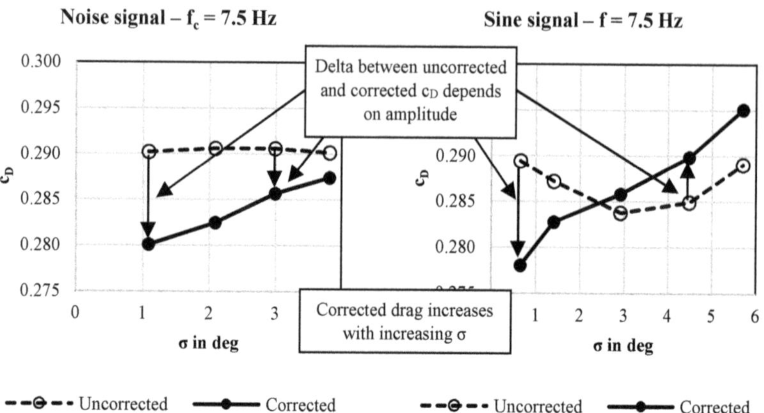

Figure 4.20:   Drag development of the DrivAer notchback model over signal standard deviation, uncorrected vs corrected, for noise signals (left) and sine signals (right)

It is evident that for both noise and sine signals the development of drag over signal standard deviation is very different between corrected and uncorrected values. While the uncorrected drag stays roughly constant with increasing signal standard deviation, the corrected drag increases. This drag increase is expected since steady-state yaw sweeps have shown that increasing flow angle generally increase drag, like seen in **Figure 4.19**. This shows that uncorrected drag results do not represent reality and should not be considered during experiments under unsteady flow.

The reason that interference correction restores the expected drag behavior with increasing yaw magnitude lies in the different pressure distributions from different flow signals discussed in section 4.2.1. The correction eliminates these unwanted pressure distribution influences on the measured drag. Therefore, the corrected c_D reflects the influence of the transient incident flow on the vehicle itself more accurately. While the interference correction also corrects for other effects like nozzle blockage and jet expansion (see section 2.4.2), those effects are mostly correlated to wind tunnel and vehicle geometry and therefore are largely the same for all unsteady flow signals. Generally, the change in pressure gradient interference caused by different flow signals is an order of magnitude higher than the change in all other interference effects caused by different flow signals.

All in all, neglecting the application of a c_D correction means measuring not only the influence of the signal on the vehicle, but also its interaction with the wind tunnel. Omitting interference corrections means that not only absolute drag results, but also effects of vehicle configuration changes will be misrepresented compared to the on-road case. Therefore, throughout this work all drag results shown have been corrected with the previously presented correction method in section 4.2.2 unless otherwise stated.

4.3 Unsteady Drag Results

The previous sections presented the wind tunnel experimental setup in the MWK, validated the correction method and proved the necessity of interference corrections. To continue, the following section will show the measured unsteady drag results for selected vehicle models. The diagrams show corrected drag plotted over signal crossflow strength, which is denoted by the signal standard deviation σ. For each vehicle model, noise and sine signals are plotted together in the same diagram. The rest of the plots and data can be found in Appendix A3.

First, the results for the DrivAer notchback model are shown in **Figure 4.21**. As mentioned previously, it is modeled after a typical sedan one might encounter on the road, and therefore is fairly representative for a common passenger vehicle.

As expected, the aerodynamic drag increases with increasing standard deviation. This is expected, as in most cases, the aerodynamic drag of a passenger car does increase when exposed to steady-state side wind (e.g. **Figure 4.19**).

Interestingly, the measured drag does not seem to correlate with the signal's frequency content at all. For both the noise and sine signals, the drag values stay in a corridor of roughly 3 counts for the same signal standard deviation. Even the type of signal – noise or sine – does not influence the c_D value as long as the signal's standard deviation remains the same.

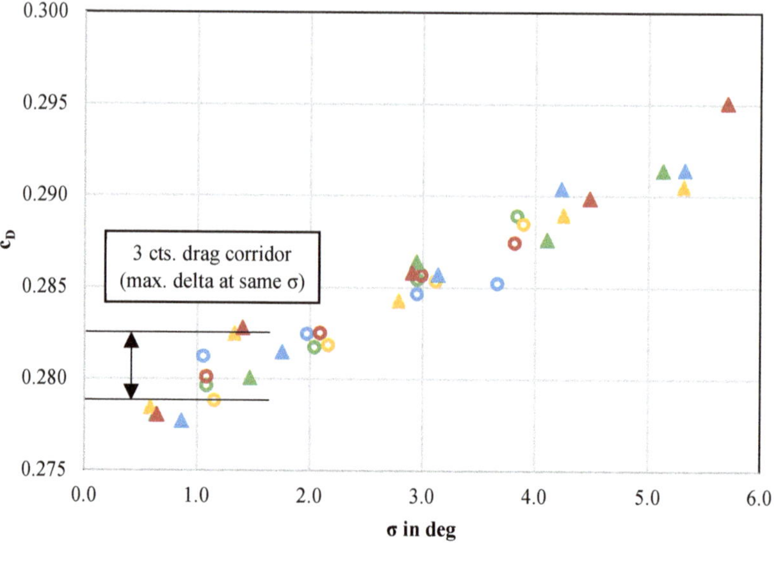

Figure 4.21: c_D values for the DrivAer notchback over standard deviation σ. At the same crossflow strength (equal σ), drag stays in a corridor of ~3 cts no matter the signals' frequency composition

Next, the quasi-steady c_D values are added to the analysis. For every measured data point, a corresponding quasi-steady drag has been determined (method: see section 3.2.1). Those are added to the data of **Figure 4.21** above, which results in **Figure 4.22** below.

The quasi-steady drag values, which have been calculated from steady-state measurements, place squarely within the drag corridor of roughly 3 counts established before. This means that, for the DrivAer notchback, the signal frequency does not influence aerodynamic drag. For this vehicle, measurements with transient yaw are superfluous, since the same results could theoretically be obtained by steady-state measurements.

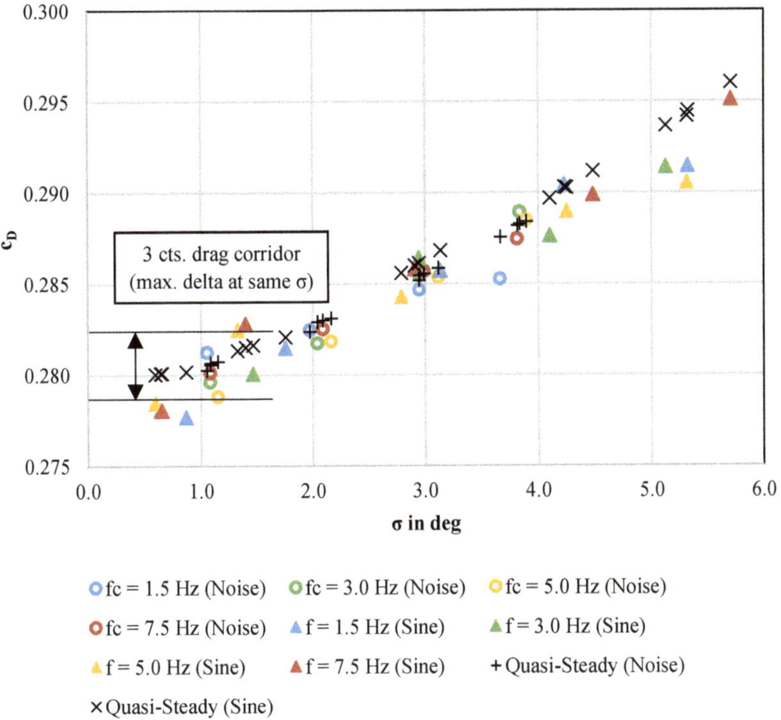

Figure 4.22: c_D values for the DrivAer notchback over standard deviation σ, with quasi-steady drag. Same 3 cts. drag corridor as before

However, it is important to not stop here but instead to also investigate other vehicles in the same manner. In doing so, it becomes clear that the conclusions made from the DrivAer notchback model are not universal and that the topic of unsteady drag is not as simple as it first might have seemed. For the SAE Body, with results shown in **Figure 4.23**, the c_D values measured for different sine signals differ significantly for the same signal standard deviation, at σ = 3° and above. Compared to the 3 cts. drag corridor for the DrivAer notchback, the data for the SAE Body show differences of up to 13 cts. at the same signal standard deviation.

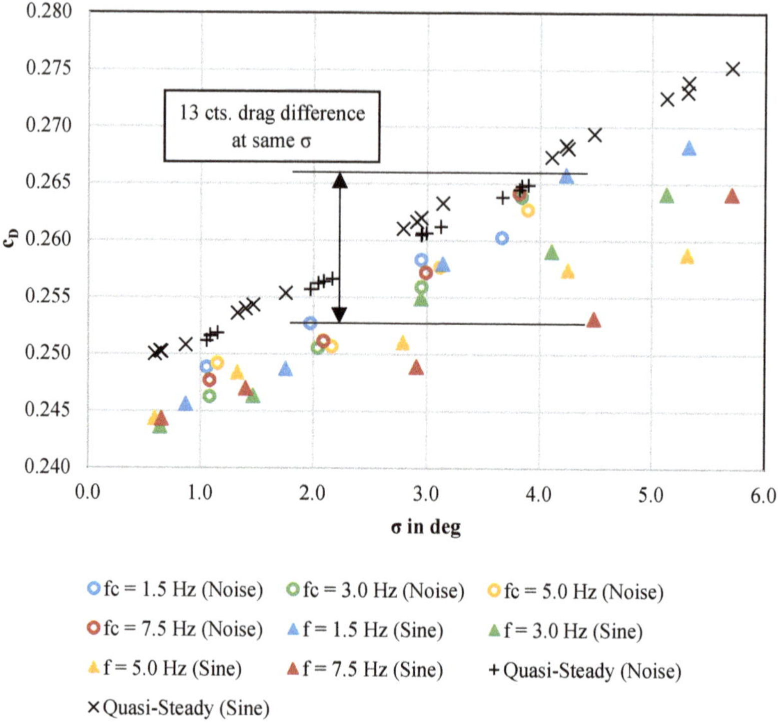

Figure 4.23: c_D values for the SAE Body in squareback configuration over σ, with quasi-steady drag. Higher frequencies result in reduced drag values compared to lower frequencies or quasi-steady flow. At the same crossflow strength (equal σ), different signals can result in up to 13 cts of drag difference

There seems to be a clear correlation between frequency and drag for the sine signals – the bigger the frequency, the smaller the drag. Furthermore, the quasi-steady c_D values are always higher than all unsteady drag values. These results prove that the signal frequency can, in fact, affect the aerodynamic drag. Additionally, for vehicles such as the SAE squareback, quasi-steady investigations are not sufficient to determine actual unsteady drag.

To summarize up to this point, the test results show that aerodynamic drag increases with increasing signal yaw angles for all test vehicles (rest of the results can be found in Appendix A3). The effect of the signal frequency, on the other hand, strongly depends on the vehicle model tested. Some vehicles, like the DrivAer notchback, barely react to frequency spectrum changes, while others, like the SAE Body in squareback configuration, react strongly.

The results for the SAE Body in **Figure 4.23** may give the impression that when signal frequency affects vehicle drag, it reduces drag the higher it is. Measurements with the AeroSUV squareback model (**Figure 4.24**), however, show this to not be the case. In the case of the AeroSUV, higher frequencies lead to higher drag compared to lower drag for the SAE Body. Furthermore, for the AeroSUV squareback, unsteady drag is mostly higher than quasi-steady drag, while for the SAE squareback it is the opposite.

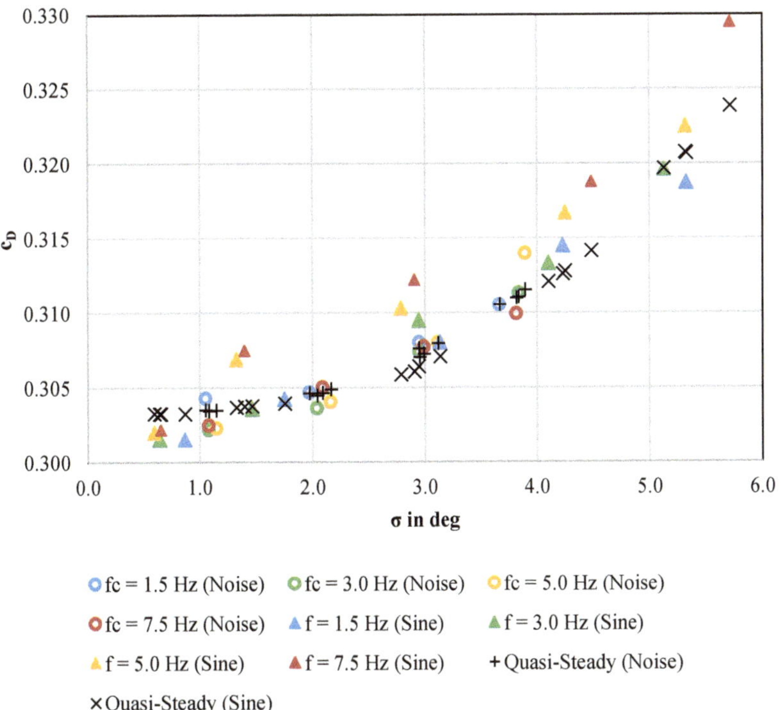

Figure 4.24: c_D values for the AeroSUV squareback over σ, with quasi-steady drag. Higher drag at high frequencies (e.g. 7.5 Hz sine, red triangles) than at low frequencies (e.g. 1.5 Hz sine, blue triangles)

In conclusion, while it was proven that some vehicle models react more strongly to signal frequency changes than others, it is not possible to establish a general correlation between frequency and drag increase or decrease. This result furthermore demonstrates the frequent non-linearity in automobile aerodynamics, which is amplified by the unsteady nature of the investigations. Nevertheless, the next section will outline ways to quantify and explain the qualitative observations made above.

4.4   New Quantifiers for Drag Reactions

The previous section has established two main findings:

1.   There is a positive correlation between signal amplitude and aerodynamic drag
2.   Signal frequency can strongly influence aerodynamic drag, but the magnitude of the influence depends on the vehicle's shape

The question now becomes what vehicle properties actually create frequency dependencies and by what amount. To learn this, it is necessary to quantify the measured drag reactions towards unsteady flow. Two different quantifiers are introduced:

1.   Drag Sensitivity Value S_D, describing the magnitude of drag change caused by frequency changes
2.   Average Delta to Quasi-Steady $\Delta_{QS,avg}$, quantifying the feasibility of approximating unsteady drag with quasi-steady drag

4.4.1   Drag Sensitivity Value S_D

To quantify a vehicle's drag reaction to changing signal frequencies, the "drag sensitivity" S_D is introduced here. S_D should be high if its corrected c_D values for the same σ value show a high difference, and low if they show a small difference. Furthermore, single data points far apart from the rest should be weighted more heavily: If one specific frequency causes a large drag variation, it implies a stronger reaction to unsteady flow than many different frequencies that only change the drag by a small amount.

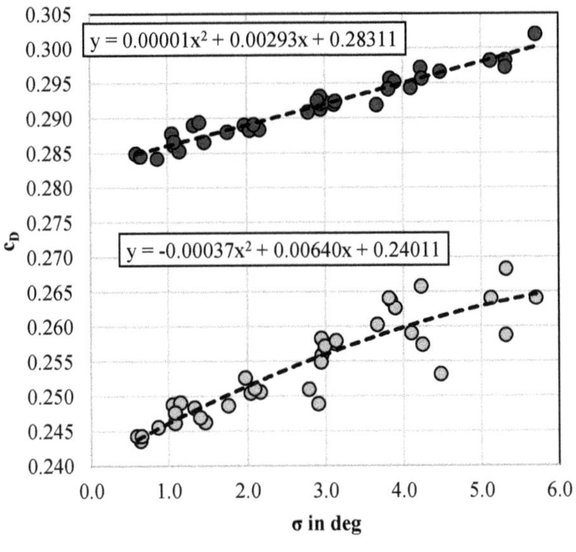

Figure 4.25: c_D values for the DrivAer notchback and the SAE Body in squareback configuration for all measured signals, with second order polynomial fitting curves

Using a least squares function approximation to define "sensitivity" will satisfy the above requirements. A second order polynomial has been chosen as the fitting curve. The fitting curves are shown in **Figure 4.25**, along with their quadratic formulae. It is now possible to calculate the residual sum of squares (RSS) for one set of data,

$$RSS = \sum_{i=1}^{n} \left[c_{D,i} - c_{D,approx}(\sigma_i) \right]^2 \qquad \text{Eq. 4.3}$$

wherein the subscript i denotes one specific data point, with a specific (cutoff) frequency and standard deviation.

While the RSS already fulfills the requirements for the "sensitivity", it cannot be easily correlated with typical aerodynamic values, and it will vary greatly with the number of data points in each data set. Therefore, "sensitivity" is instead defined as follows.

$$S_D = \sqrt{\frac{1}{n} * \text{RSS}} \qquad\qquad \text{Eq. 4.4}$$

Here, the term under the square root is the average square of errors for the data set. This way, S_D provides a first order statistic, unlike RSS which is a second order one, making S_D easier to relate to c_D numbers. Thus, "sensitivity" represents a weighted average offset of the data set's c_D values from the fitting curve. If, for instance, the sensitivity for a certain vehicle model to a certain group of unsteady signals is 0.003, or 3 counts, this means that on average, the measured c_D will deviate from the fitting curve by 3 counts.

A high drag sensitivity vehicle will experience a large change in drag on signal frequency changes at constant signal amplitude, while a low drag sensitivity vehicle will have little change in drag at constant signal amplitude despite changes in frequency. With the data from **Figure 4.25**, the DrivAer notchback has a sensitivity of 1.0 counts while the SAE squareback has a sensitivity of 2.9 counts, confirming the qualitative observation that the SAE squareback reacts more strongly to frequency changes than the DrivAer notchback.

The choice of a second order polynomial fitting curve stems from the steady-state yaw sweeps measured in this work that showed the strongest non-linear behavior. **Figure 4.26** shows the yaw sweep results for the DrivAer and Aero-SUV squareback models that suggest at least a second order polynomial.

Figure 4.27 shows the drag values over standard deviation for the same models and also supports the suitability of a second order approximation curve. Choosing a higher order fitting curve would only change S_D values by a magnitude of ~0.1 counts compared to the second order curve. Therefore, the second order curve was ultimately chosen.

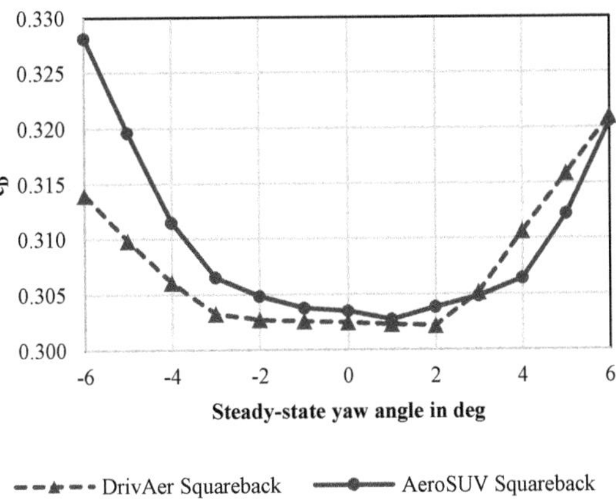

Figure 4.26: c_D results for a steady-state yaw sweep of the DrivAer and AeroSUV squareback, showing second order behavior

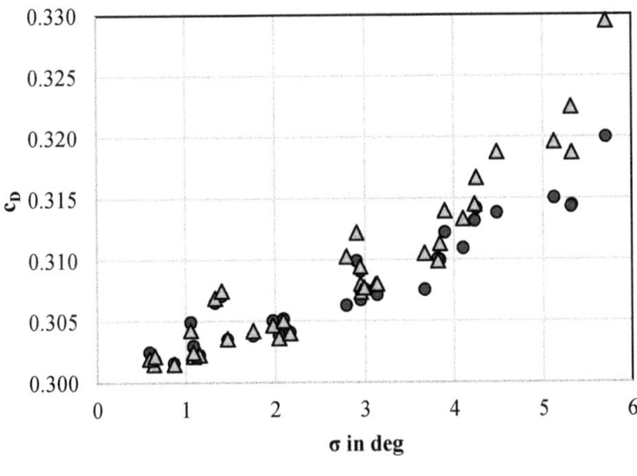

Figure 4.27: c_D results for the DrivAer and AeroSUV squareback under all unsteady flow signals, showing second order behavior

4.4.2   Average Delta to Quasi-Steady $\Delta_{QS,avg}$

A second way to quantify the effect of unsteady flow on drag is comparing the delta between unsteady and quasi-steady drag. By calculating the difference between the measured unsteady c_D and its respective quasi-steady c_D, it is possible to evaluate the influence of the unsteady flow aside from the influence of nonzero yaw.

Figure 4.28 shows the same selection of results as **Figure 4.25**, but with the quasi-steady drag values also plotted.

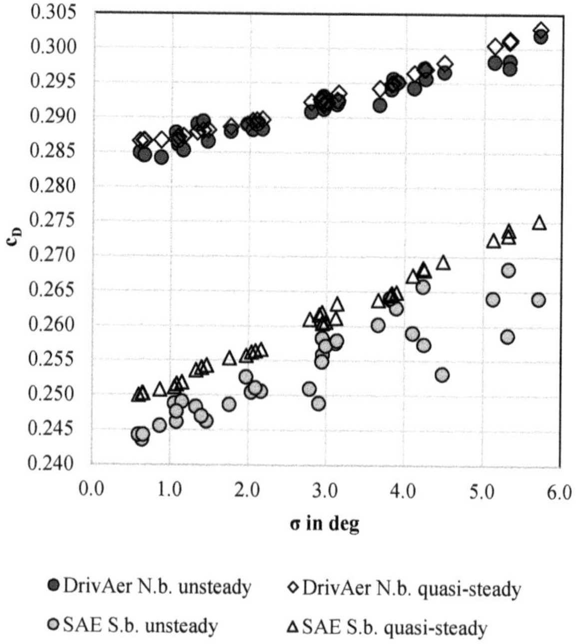

Figure 4.28:   Corrected c_D values for the DrivAer notchback model and the SAE squareback model, with quasi-steady drag

Now it is possible to calculate the sum of the differences between the unsteady c_D and the quasi-steady c_D for each data point i.

$$\text{Sum}(\Delta^2) = \sum_{i=1}^{n} \left(c_{D,i} - c_{D,quasi-steady,i} \right)^2 \qquad \text{Eq. 4.5}$$

Just like during the calculation of drag sensitivity described previously in section 4.4.1, the difference is squared first in order to weigh the sum towards the higher differences. The final quantifier is then defined using $\text{Sum}(\Delta^2)$ like drag sensitivity is defined using RSS.

$$\Delta_{QS,avg} = \sqrt{\frac{1}{n} * \text{Sum}(\Delta^2)} \qquad \text{Eq. 4.6}$$

"$\Delta_{QS,avg}$" is the abbreviation for "average delta to quasi-steady". The higher $\Delta_{QS,avg}$ is for one vehicle model, the higher the difference is between its measured drag values and its calculated quasi-steady drag values on average.

As mentioned in section 4.3, the drag increase a vehicle experiences under unsteady flow is caused by both steady yaw effects and unsteady frequency effects. $\Delta_{QS,avg}$ quantifies the latter, with higher values corresponding to larger frequency effects. The DrivAer notchback in the above example (**Figure 4.28**) has an $\Delta_{QS,avg}$ of 1.5, while the SAE squareback has an $\Delta_{QS,avg}$ of 7.1, confirming that frequency effects on drag are much stronger on the SAE squareback model.

Though both S_D and $\Delta_{QS,avg}$ ultimately quantify the unsteady influence on vehicle drag, it is important to note that the two values are different and useful in different cases. S_D is more useful when trying to judge a vehicle model's drag reactions to a set of signals with varying frequency spectra, irrespective of how far these particular signals move the drag value off the steady-state drag. In contrast, $\Delta_{QS,avg}$ is more useful in determining whether the vehicle model is influenced at all by any unsteady flow signals. If $\Delta_{QS,avg}$ is very low for a number of signals close to the actual use case, like on the road, it is even possible to forgo unsteady measurements and replace them with the quasi-steady measurements for a similarly accurate prediction of unsteady drag, as described for the DrivAer notchback in section 4.3.

Finally, it is vital to consider that sensitivity and $\Delta_{QS,avg}$ are dependent on both the vehicle model and the unsteady flow signal used in the experiments. For the purposes of the present research, however, they are dependent only on the model because all models have been measured with the same signals. Since the real, on-road flow is well approximated by the noise signals in this paper, the relative sensitivity and $\Delta_{QS,avg}$ between the different models can nonetheless be considered representative of their geometric differences.

4.5   Quantifier Results

4.5.1   Cumulative Results from All Signals

Figure 4.29 shows the Sensitivity and Delta to Quasi-Steady values for each model, each calculated from a data set with measurements for all 36 unsteady excitation signals.

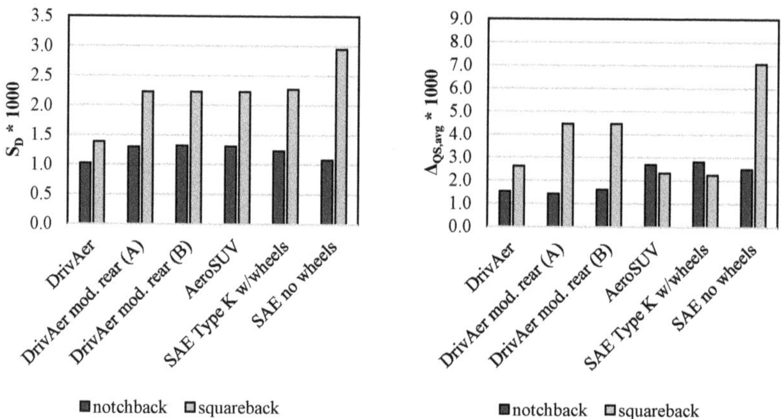

Figure 4.29: S_D and $\Delta_{QS,avg}$ results for each measured vehicle model with all unsteady signals (16 noise, 20 sine) [74]

Generally, the following geometric traits increase S_D and $\Delta_{QS,avg}$:

1.   Squareback geometry as opposed to notchback geometry
2.   An increased rear separation area, as evidenced when comparing the DrivAer to the AeroSUV

3. A simple, angular rear geometry, like on the SAE Body. This can be observed when comparing the standard DrivAer to the modified DrivAer variants with sharp-edged rear ends
4. The presence of wheels for the SAE notchback, but the lack of wheels for the SAE squareback

Traits 1-3 relate to the rear end wake of the vehicle, which plays a considerable role in its aerodynamic drag. The above observations suggest that larger wakes created by traits 1 and 2 are more prone to change when beset with unsteady incident flow. Trait 3 (sharp rear ends) likely creates clear cut, coherent wake structures that have the same effect. Trait 4 likely stems from wholly different interactions of the SAE Type K's wheels with the notchback and squareback rear end shapes. Possibly, these interactions stem from the change of the separation area's shape due to the wheels or the sharp-edged wheelhouses.

However, traits 1 and 2 are not valid for the squareback AeroSUV and SAE Type K models with respect to $\Delta_{QS,avg}$. Both models do not measure larger $\Delta_{QS,avg}$ than their notchback counterparts, nor does the larger AeroSUV squareback measure a larger $\Delta_{QS,avg}$ compared to its DrivAer counterpart. A possible explanation for the AeroSUV's results is its use of the exact same rear end attachments as the DrivAer. Since the AeroSUV is larger than the DrivAer, changing the rear end attachment causes a comparatively smaller change in geometry in the AeroSUV compared to the DrivAer, which could lead to different reactions in unsteady drag. For the SAE Type K, the difference to the other models again likely comes down to different wake interactions of the wheels and wheelhouses with the two rear end shapes.

4.5.2 Separated Results by Noise and Sine Signals

By separating the results from the noise and sine signals and evaluating those separately, it is possible to learn about the specific influence of the different signal types on aerodynamic drag. As previously mentioned (section 4.1.2), the noise signals represent realistic unsteady flow fields, while the sine signals' simplicity and regular time history allow easier examination of its effects on drag. **Figure 4.30** shows the Sensitivity and Delta to Quasi-Steady values calculated from the measurements of the 16 noise signals, while **Figure 4.31** shows the same values calculated from the measurements of the 20 sine signals.

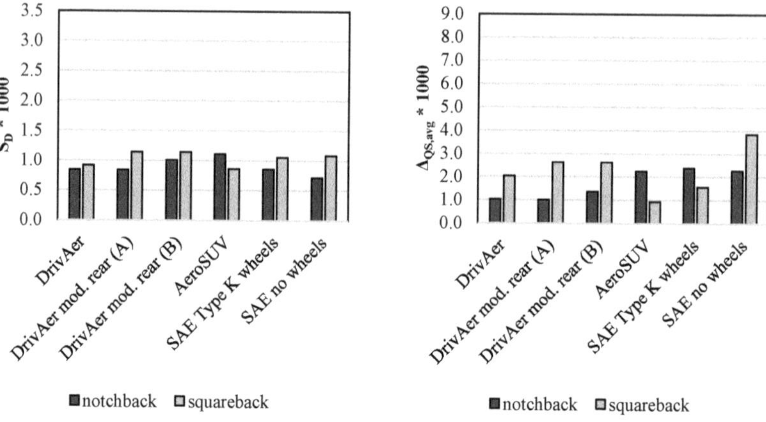

Figure 4.30: S_D and $\Delta_{QS,avg}$ results for each measured vehicle model with the 16 noise signals [74]

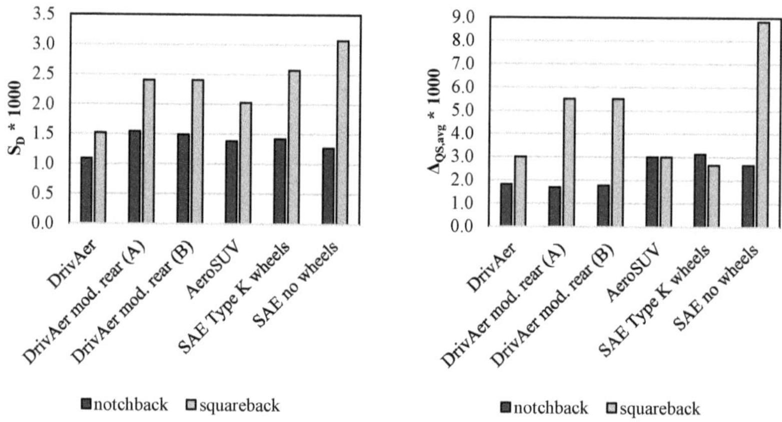

Figure 4.31: S_D and $\Delta_{QS,avg}$ results for each measured vehicle model with the 20 sine signals [74]

As evident in **Figure 4.30** and **Figure 4.31**, both Drag Sensitivity and Average Delta to Quasi-Steady values are lower across the board for the noise signals compared to the combined results (section 4.5.1). For the sine signals, it is the opposite: Drag Sensitivity and Average Delta to Quasi-Steady are higher compared to the combined results. The possible explanations for this are as follows:

1. The applied noise signals' standard deviation values, and therefore their incident flow yaw angles, are generally lower than the sine signals'. The maximum standard deviation for the noise signals is 3.9°, while it is 5.7° for the sine signals. A lower deflection leads to a lower dispersion of c_D results, which reduces S_D and $\Delta_{QS,avg}$.
2. For a noise signal, the most common yaw angles are around 0°, which reduces yaw effects. For a sine signal, the most common yaw angles are close to the maximum yaw angle, which increases yaw effects.
3. Each noise signal contains multiple frequencies compared to the sine signal's single one. The different cutoff frequencies f_c merely removes all frequencies above it. Therefore, noise signals with different f_c are still somewhat similar. They will share the low frequencies and therefore may cause similar effects on vehicle drag, reducing the calculated unsteady quantifier values.

The observations in the previous section on the geometric traits that increase S_D and $\Delta_{QS,avg}$ also generally apply here,

1. Squareback geometry as opposed to notchback geometry
2. An increased rear separation area
3. A simple, angular rear geometry, like on the SAE Body
4. The presence of wheels for the SAE notchback, but the lack of wheels for the SAE squareback

as well as the AeroSUV and SAE Type K models behaving counter to traits 1 and 2.

The trends are the same compared to the cumulative results in section 4.5.1. However, the quantitative observations in the separated results are clearer, i.e. have higher magnitudes, when looking at the sine signals' results. As mentioned, S_D and $\Delta_{QS,avg}$ resulting from sine signals are higher across the board, providing greater differentiation. Consequently, sine signals will be useful in the future despite a lack of similarity to on-road flow. The effects of changes in frequency and amplitude can be easily tested using sine signals and then validated using more realistic signals, if necessary.

4.6   Summary

The experimental measurements conducted in the Model Scale Wind Tunnel led to the following results:

1. Different unsteady flow signals result in different static pressure distributions in the test section that have to be accounted for. Therefore, wind tunnel interference corrections are absolutely necessary for unsteady drag investigations using a turbulence generation system like FKFS swing or similar. Not applying corrections does not just result in an offset error like with steady-state measurements, but also in tendency errors that lead to wholly wrong conclusions.

2. Unsteady aerodynamic drag increases with increasing signal standard deviation / yaw magnitude. This is equivalent to drag increasing with yaw angle in a steady state yaw sweep and is not caused by flow transience.

3. Changes in signal frequency, but not in yaw angle, affect drag in different ways depending on vehicle geometry. Some models are very sensitive to frequency changes, others not at all. No general correlations could be found between signal frequency and drag.

4. Two new unsteady metrics S_D and $\Delta_{QS,avg}$ were introduced to quantify the frequency effects on drag. A large S_D indicates a high magnitude of drag change caused by frequency changes. A high $\Delta_{QS,avg}$ indicates low feasibility of approximating unsteady drag with quasi-steady drag.

5. Evaluating S_D and $\Delta_{QS,avg}$ for the measured vehicle models resulted in insight about what geometric traits influenced those newly created metrics. It was concluded that highly influential traits strongly influence the vehicle's various wake sizes and shapes.

Through wind tunnel measurements, the effects of unsteady flow on drag were successfully quantified and geometry correlations established. What remains, however, is the question of how similar such results obtained through a dynamic turbulence generation system actually are compared to on-road flow and drag. To answer these questions, CFD offers an ideal tool, as it is able to simulate both the open road as well as the wind tunnel environment.

5 CFD Investigations and Results

Previously, chapter 4 described the experimental investigations in the wind tunnel made with a multitude of different unsteady flow signals and vehicle geometries. Two major conclusions can be made from the results:

1. When using a dynamic turbulence generation system like FKFS swing in the wind tunnel, interference effects are inevitable and need to be considered and/or corrected. A correction method for unsteady measurements has thus been established and validated.
2. Incident flow frequency influences drag on different vehicles differently depending on various geometric traits. These influences have been quantified.

The numerical investigations detailed in the current chapter build upon those two conclusions and will attempt to answer the resulting questions:

1. How accurate are measurements with FKFS swing compared to blockage-free on-road drag values? How big are the differences and what are their causes?
2. How do different signal frequencies result in different unsteady drag in vehicles with high drag sensitivity?

The present chapter will first outline the numerical setup, including the overall simulation environment, wind tunnel configuration, vehicle and signals, and chosen measurements. Next, the validity of the simulation setup will be discussed. Finally, the CFD results will be presented and discussed, including measured drag, flow fields, and unsteady flow analyses.

5.1 Numerical Simulation Environment

The CFD software Simulia PowerFLOW, version 6-2019-R4, was used throughout all investigations. Section 2.5 goes into more detail about the basics of CFD and discretization. Specifically, PowerFLOW discretizes the simulation volume using cubic cells called "voxels" and different levels of varia-

© The Author(s), under exclusive license to
Springer Fachmedien Wiesbaden GmbH, part of Springer Nature 2026
X. Fei, *The Impact of Unsteady Flow on Drag Measurements in Automotive Wind Tunnels*, Wissenschaftliche Reihe Fahrzeugtechnik Universität Stuttgart,
https://doi.org/10.1007/978-3-658-51766-3_5

ble resolution, called "VR regions". In the simulations described here, the finest cells are defined as VR level 9, with a width of 0.6 mm. Each subsequently coarser VR region doubles in cell size. Therefore, VR8 voxels are 1.2 mm wide, VR7 voxels 2.4 mm, and so on.

Out of the vehicle models described in section 4.1.3, the DrivAer squareback model was chosen as the focus because it is well validated in PowerFLOW, especially for unsteady incident flow cases [64]. Compared to the notchback, it is more sensitive towards changes in unsteady flow (see section 4.5), and thus better suited to evaluate the changes unsteady flow creates. To ensure comparability with the measurements, neither ground movement nor engine bay flow were simulated, and the simulation model remained in 1:4 scale.

As for the unsteady flow signals, sine signals were chosen because they only contain one discrete frequency. This makes it easier to correlate that frequency with resulting frequencies in the flow around the vehicle. Additionally, sine signals have short periods that easily fit in realistic simulation timeframes of < 10 seconds. A more in-depth discussion on these points can be found in section 4.5.2. The table below shows the sine signals used in the CFD simulations.

Table 5.1:   Sine signals used during CFD investigations

f in Hz	St @ 45 m/s, 1:4 scale	A_{flap} in °	σ_{flap} in °
1.5	0.038	5.00	3.54
3.0	0.077	2.50	1.77
3.0	0.077	5.00	3.54
5.0	0.128	5.00	3.54
7.5	0.192	2.50	1.77
7.5	0.192	5.00	3.54

A nominal sine amplitude of 5° with all four frequencies was chosen because the large flow angles result in greater differences in the results, easing observations. At the same time, the amplitude is small enough to keep its influence on the static pressure distribution low (see section 4.2.1). Additionally, two further signals at an amplitude of 2.5° have been added to investigate possible changes in the results with lower amplitude.

Two different overall simulation setups were utilized and will be explained in the next sections. The Digital Model Scale Wind Tunnel (DMWK) is a detailed geometric representation of the IFS Model Scale Wind Tunnel. The Digital Wind Tunnel (DWT), on the other hand, is a conventional blockage-free simulation environment without additional geometries, but adapted to unsteady flow.

5.1.1 Digital Model Scale Wind Tunnel (DMWK)

The DMWK was first created by Fischer [66] [65] [67] in order to compare wind tunnel measurement results with CFD simulations and has been continuously updated since then. The current setup is based on the work of Stoll [64], who adopted it in order to use it with the swing system. Details not mentioned in this section, such as specific discretization details, can be found in the aforementioned publications.

Instead of providing just a simple box environment, the DMWK includes the geometries of the Model Scale Wind Tunnel's settling chamber, nozzle, test section, plenum, collector and diffuser. Therefore, it closely mimics a digital recreation of wind tunnel experiments.

As in the experiments, the CFD simulations operate with a freestream velocity of 45 m/s (162 km/h). In the DMWK, this is achieved using mass flow boundary conditions. The inlet and outlet are located at the start of the settling chamber and the end of the diffuser, respectively. The mass flow is set so that in the middle of the empty test section the flow velocity is 45 m/s. The center belt and wheel drive units are also present in the test section and the center belt is equipped with its physical counterpart's measured roughness. This way, the floor boundary layer matches the experiment.

Like in the physical wind tunnel, the unsteady incident flow in the test section is created by the FKFS swing gust generation system at the nozzle exit (**Figure 5.1**). Initially, the flow coming from the mass flow inlet and entering the nozzle is steady and uniform. Upon reaching the rotating airfoils, it is deflected according to the momentary airfoil position. In order to create accurate comparisons, the flap angles chosen for the unsteady signals are the same as the ones used in the experiments. Simulations with an empty test section have shown that the resultant flow angles in the center of the test section match those measured in the physical wind tunnel.

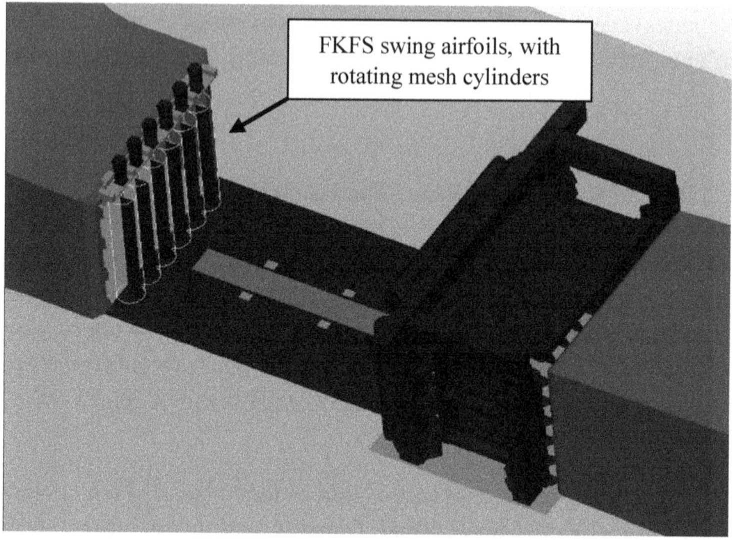

Figure 5.1:   DMWK test section, with FKFS swing

Cylindrical sliding mesh sections are used to model FKFS swing's dynamic airfoil motion, which can be seen in **Figure 5.1**.

The VR regions are chosen so that the test section, the shear layer, the floor boundary layer and the airfoil motion are represented accurately. Power-FLOW's Best Practice setup [61] is largely followed regarding the vehicle model. The VR6 and VR7 regions are an exception, as they widen out towards the back by 10°, as in Stoll's investigations [64], in order to capture transient effects at the model's sides (**Figure 5.2**)

Henceforth, the DMWK environment will also be referred to as "simulated wind tunnel environment" to better differentiate with the Digital Wind Tunnel outlined in the next section.

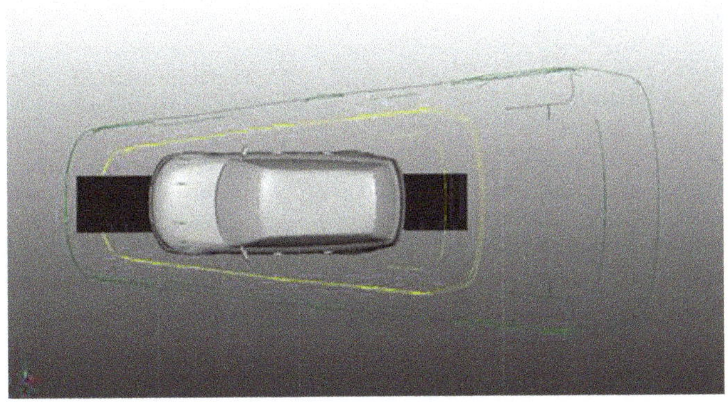

Figure 5.2:   Top view of DrivAer; VR6 (green) and VR7 (yellow) regions

5.1.2   Digital Wind Tunnel (DWT)

The Digital Wind Tunnel represents an idealized on-road case without any obstructions or interference effects, and is therefore a predictor of on-road aerodynamic performance. It consists of a typical box setup much larger than the model, with a blockage ratio of 0.1 %.

The box dimensions and VR regions, including the vehicle ones, were also chosen according to PowerFLOW Best Practice [61]. However, some key differences exist compared to a steady-state flow setup. First of all, the wind tunnel walls were removed so that flow in y-direction does not interact with them. Secondly, the model's position is much farther forward than in a steady-state setup (**Figure 5.3**) like in D'Hooge's investigations [60]. This is necessary because otherwise the longer distance between inlet and model will dampen the transient flow and therefore reduce simulation accuracy. For the same reason, all rectangular VR boxes (VR1 − VR5) have been extended forward to the velocity inlet, like in Jessing's CFD investigations [4].

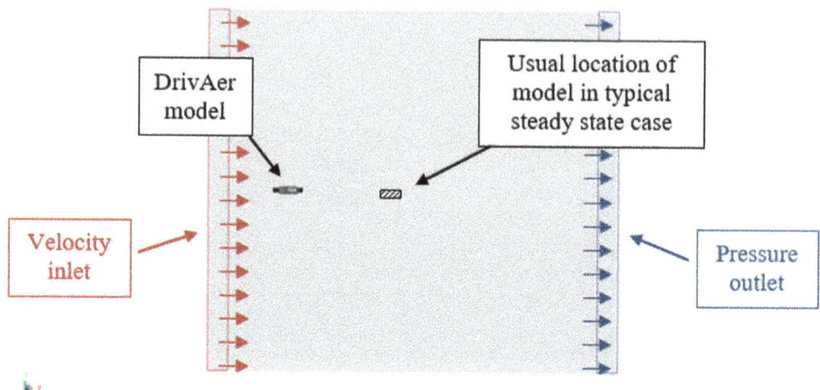

Figure 5.3:   DWT wind tunnel setup, showing the forward vehicle position
for all simulations. The usual location of the model for typical
steady state simulations is also shown for comparison

Compared to the simulated wind tunnel environment, the DWT also retains
the modified, divergent VR6 and VR7 regions shown in **Figure 5.2**. The floor
layout with the center belt and wheel drive unit also remains the same to create
the same floor boundary layer structure.

The unsteady flow in the DWT is created using a time-resolved velocity
boundary condition at the inlet. The x- and y-velocity are chosen so that the
desired flow angle is achieved and that the velocity magnitude remains at
45 m/s (162 km/h). The desired flow angle is determined by the chosen sine
signal. Its amplitude is the test section flow amplitude obtained by simulating
the signal at its flap amplitude (5° or 2.5°) in the DMWK. Equations 5.1 and
5.2 show this in mathematical terms.

$$\sqrt{v_x^2 + v_y^2} = v_\infty = 45 \; m/s \qquad \text{Eq. 5.1}$$

$$v_y = A_{DMWK} \sin(2\pi f t) \qquad \text{Eq. 5.2}$$

Henceforth, the DWT environment will also be referred to as "free flow environment" to better differentiate with the simulated wind tunnel environment
described in section 5.1.1.

5.1.3 Measurements

All unsteady simulations run for 9 seconds, of which the timeframe from 3 to 9 seconds are evaluated. The start at 3 seconds allows the flow to settle into a regular pattern. The 6 seconds of evaluation time is a multiple of the period for all used sine signal frequencies, ensuring consistent evaluations. Steady-state simulations run for 5 seconds and are evaluated between 2 to 5 seconds to save computational resources.

Most measurements are averaged over the evaluation time and then analyzed. Among those are force/moment coefficients, static pressure distributions, boundary layer distributions and qualitative flow field images. In order to evaluate unsteady flow effects, multiple virtual flow probes have been placed around the model. Those probes capture flow values at 1250 Hz, the same frequency as the cobra probe in the real wind tunnel (see section 4.1.1). For the following analyses, the important time-resolved flow values captured are:

- Velocity
- Flow angles (resultant from velocity components)
- Static and total pressure
- Vorticity

These values also allow the calculation of further unsteady values such as turbulence intensity and length scale.

Like the experimental drag results in the Model Scale Wind Tunnel, the simulated drag results in the simulated wind tunnel have also been corrected for interference effects. For the horizontal buoyancy corrections, the gradients used were those obtained during CFD simulations with an empty test section. This is necessary because despite being close to the measured gradients, the simulated ones (presented in section 5.2.1), are not exactly the same. The rest of the input data for the corrections, such as wind tunnel and model geometry, remained the same for experiment and CFD. Henceforth, all presented simulated wind tunnel (DMWK) drag results will have been corrected for interference unless otherwise stated. Free flow environment (DWT) drag results are, by nature, correction-free.

5.2   CFD Validation

Before getting into drag and flow field evaluations, the CFD setups must first be validated. For steady-state flow, both flow environments have been used and validated by a number of researchers, such as Fischer et al. [65] [66], Stoll [40] and Jessing [4]. The addition of unsteady incident flow, however, necessitates further validation efforts. Section 5.2.1 will first validate the empty test section using MWK pressure gradient and boundary layer results. Section 5.2.2 will compare select CFD drag results with MWK measurements discussed previously in chapter 4.

5.2.1   Static Pressure Gradient and Boundary Layer

Figure 5.4 shows an exemplary pressure distribution for an unsteady flow signal. The DMWK pressure distribution conforms to the experimental one. The pressure increase at the rear of the test section caused by the unsteady flow (see section 4.2.1) is present in both experiment and CFD.

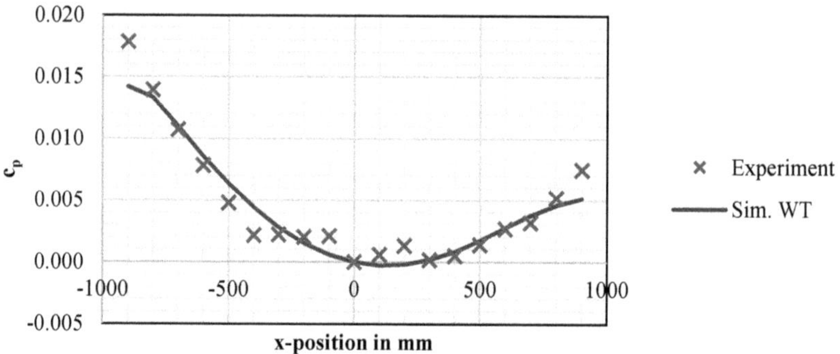

Figure 5.4:   Static pressure distribution for sine flow at f = 3 Hz and A = 5°

Figure 5.5 shows the pressure delta values (definition in eq. 4.2) in the empty test section compared to the experimental results. The trend of increasing pressure delta with changing signal frequency stays the same between experiment and CFD. The maximum absolute difference in pressure delta between experiment and CFD is 0.0027, or 2.7 counts, which is a good correlation.

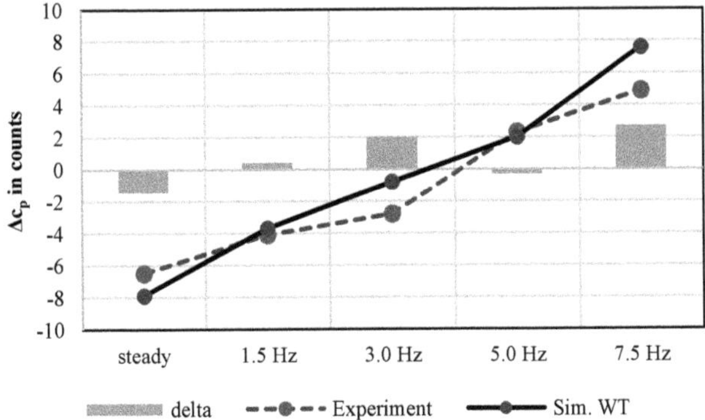

Figure 5.5: Pressure deltas front to back in the empty test section for sine signals with A = 5°; experiment and simulated wind tunnel

The boundary layers shown in **Figure 5.6** also show a high similarity between experiment and CFD, at different positions along the wind axis.

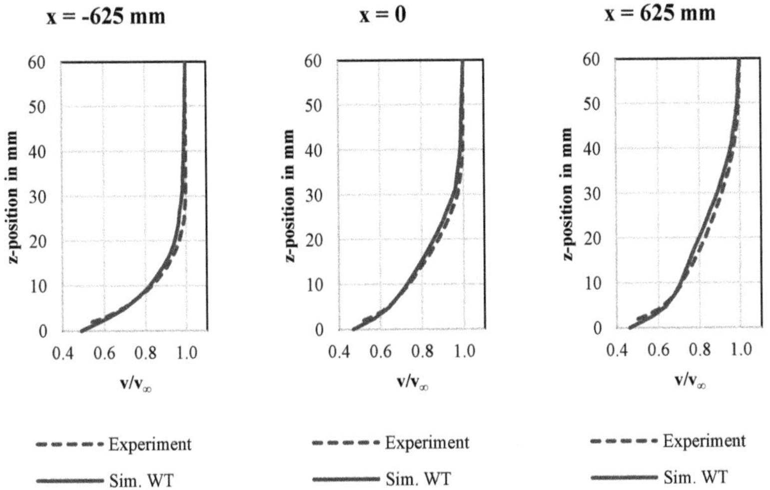

Figure 5.6: Boundary layer velocity distribution in the empty test section, above the center belt, for steady-state flow at zero yaw

Overall, the static pressure distribution and boundary layer results in the simulated wind tunnel environment are sufficiently similar to the experimental results to validate this CFD environment and setup.

For the free flow environment, no experimental comparisons are available because such a perfect blockage-free environment is impossible to reproduce in the real world. However, resultant flow angles and velocities were taken at the vehicle position using empty free flow simulations. These confirmed that both angles and velocities the vehicle experienced were equal to the ones set as the inlet boundary condition. In addition, it was confirmed that the floor boundary layer matches the one in the simulated wind tunnel setup.

5.2.2   Drag Coefficient

In order to validate the CFD simulations with the model in the test section, the overall drag results in the Model Scale Wind Tunnel can be compared to the ones obtained in the simulated wind tunnel. **Figure 5.7** shows those drag results:

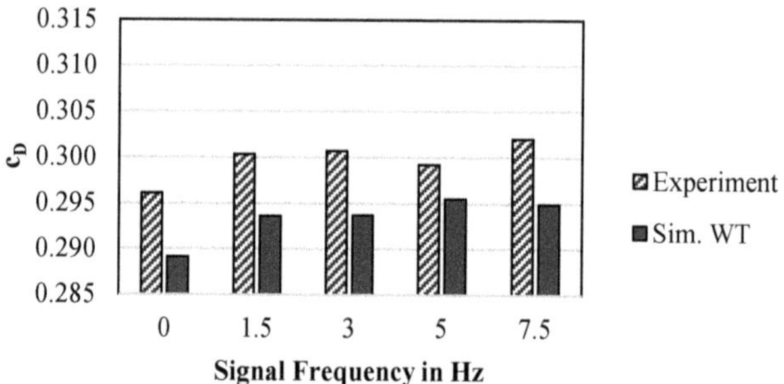

Figure 5.7:   Comparison between experimental and simulated wind tunnel DrivAer squareback c_D results, for sine signals at 5° nominal amplitude (0 Hz: steady-state flow at 0° yaw)

The difference between experiment and simulated wind tunnel is between 4 and 7 drag counts, which is an acceptable amount to establish correlation.

While the experimental and simulated drag values in the wind tunnel match up well, the same cannot be said for the drag values in the simulated wind tunnel compared to the simulated free flow (**Figure 5.8**). However, this is to be expected since the free flow environment is very different from the experimental environment.

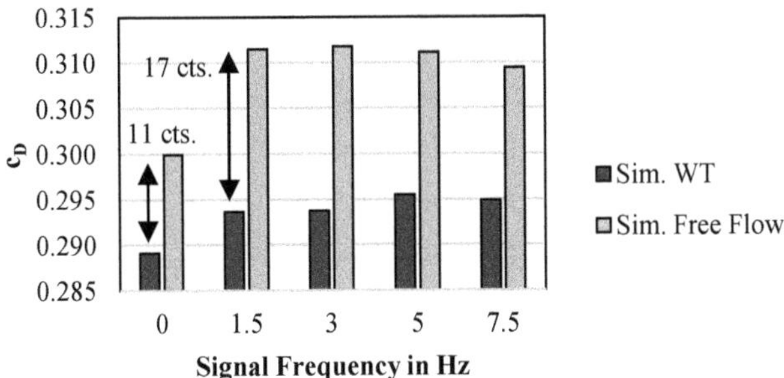

Figure 5.8:   Comparison between simulated wind tunnel and simulated free flow DrivAer squareback c_D results, for sine signals at 5° nominal amplitude (0 Hz: steady-state flow at 0° yaw)

The discrepancy is not overly problematic in itself, since aerodynamic development usually concerns itself with deltas between different configurations. However, the deltas here are not constant either. The drag delta between the environments is especially different going from steady to unsteady flow, rising from 11 counts for the former to 14–17 counts for the latter.

All in all, the CFD simulation setup has been validated using experimental measurements, showing matching pressure distributions, boundary layers and drag results. The differences between wind tunnel measurements and simulated free flow environment, while not indicative of an invalid simulation setup, will be expanded on in the next section.

5.3   Drag Force Development

In order to determine what exact areas influence drag differently in the wind tunnel and in free flow, the force development tool is used. **Figure 5.9** and **Figure 5.10** shows the drag force development diagrams for all signals at $A = 5°$ as their difference to the steady-state simulation. Surface pressure is integrated over a section of 50 mm for each data point, continuing along the vehicle length. The result at each point shows how much these 50 mm of model length contribute to total drag. Positive values mean increasing drag force compared to the steady-state case and negative values mean decreasing drag force compared to the steady-state case. All force development diagrams in this section have been obtained using data uncorrected for interference effects, because no correction methods for integrated surface pressures exist.

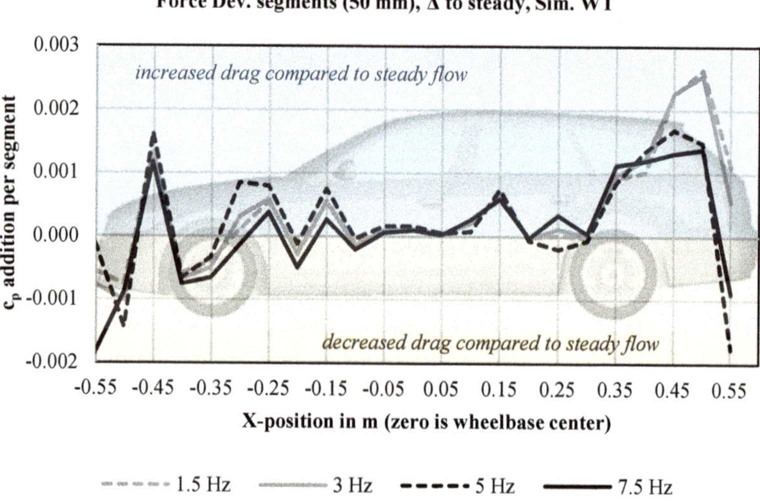

Figure 5.9:   DrivAer squareback drag force development, 50 mm segments, delta to steady-state, simulated wind tunnel environment, sine signals at A = 5°

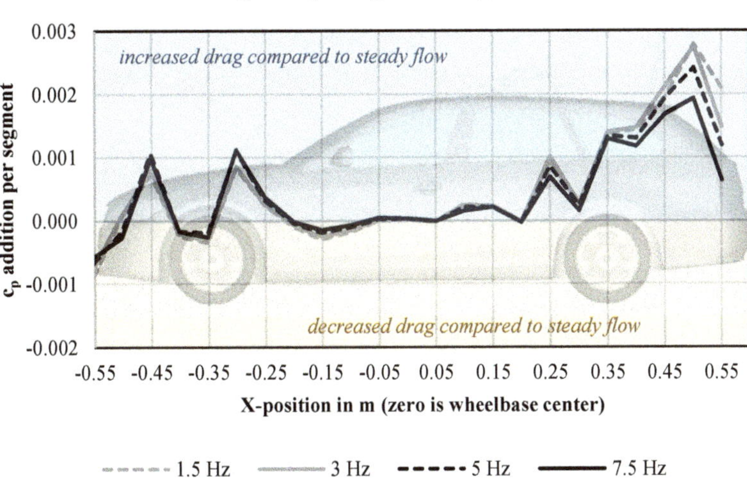

Figure 5.10: DrivAer squareback drag force development, 50 mm segments, delta to steady-state, free flow environment, sine signals at A = 5°

From these diagrams, the following observations can be made:

- The major sections causing differences between steady and unsteady drag are the (static) wheels and the rear end (both environments)
- The major sections causing differences between different signal frequencies are the front end, the front wheels and the rear end (simulated wind tunnel), and only the rear end (free flow)

Differences between the two environments are visible, but difficult to pinpoint from the aforementioned diagrams. Therefore, **Figure 5.11** further subtracts the free flow values from the simulated wind tunnel values to enable a clear comparison.

This diagram confirms the front end, front wheels and rear end as critical areas to observe, since they cause the biggest differences between the simulated wind tunnel and free flow setups.

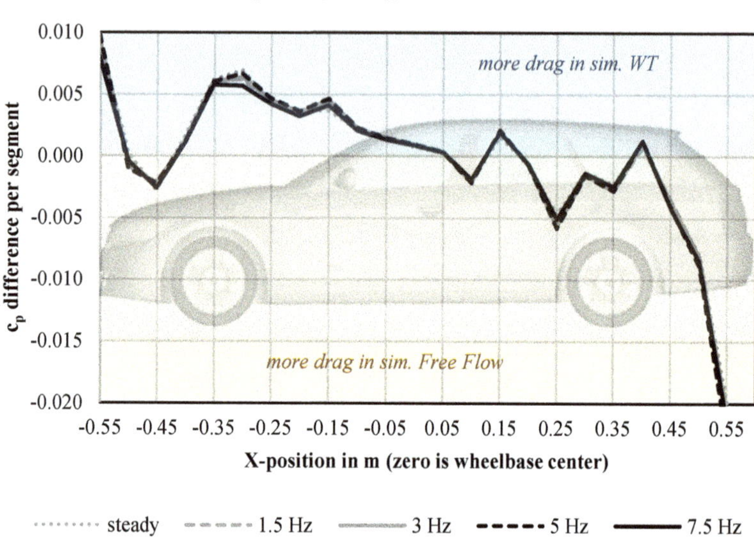

Figure 5.11:   DrivAer squareback drag force development, 50 mm segments, delta to steady-state, difference between simulated
wind tunnel and free flow environments, sine signals at A = 5°

5.4   Flow Analysis near the Vehicle

Having determined the areas of interest on and around the vehicle, the next
step is an in-depth analysis of these areas. The following zones will be investigated:

- Front end
- Rear end
- Front wheels

For all zones, the following subjects were analyzed:

- Flow fields
- Flow yaw angle development
- Average pressure and fluctuation

- Turbulence intensity and length scale
- Fast Fourier transforms for relevant transient values

In the following sections, only the results yielding progress to answering the research questions outlined in at the beginning of the chapter will be shown.

5.4.1 Front End

Section 5.3 shows a large difference in added drag between both simulation environments at the front end (**Figure 5.11**), as well as high frequency dependency in the simulated wind tunnel that does not occur in the free flow environment (**Figure 5.9**, **Figure 5.10**). These observations can all be explained with the different, non-zero static pressure distributions resulting from unsteady flow in the wind tunnel. The high static pressures at the front of the model in the simulated and real wind tunnels conform to the high pressures measured in the empty test section (see e.g. **Figure 4.12**).

However, **Figure 5.8** does show that overall drag is much higher in free flow compared to the wind tunnel environment. While this is certainly not only due to flow differences at the front end, it is still prudent to investigate possible causes here. One cause is likely the flow angle in front of the vehicle. **Figure 5.12** compares the flow angle time history of the same signal in the simulated wind tunnel and free flow environments.

The diagram shows sinusoidal flow angle time history for both environments, indicating that the model and the wind tunnel are not yet influencing the frequency distribution of the flow at this location. However, the amplitude of the sine function differs between the simulated wind tunnel and box environments, with the free flow environment providing around 10 % greater amplitudes. While flap angles in the wind tunnel do not translate to the same flow angles in the test section, care has been taken to achieve the same empty test section flow angles for both simulated wind tunnel and free flow environments (see section 5.1.2). Therefore, one key takeaway is that the presence of a vehicle in the test section changes the correlation function between flap and flow angle in the wind tunnel. This correlation function becomes vehicle-dependent and is no longer universal.

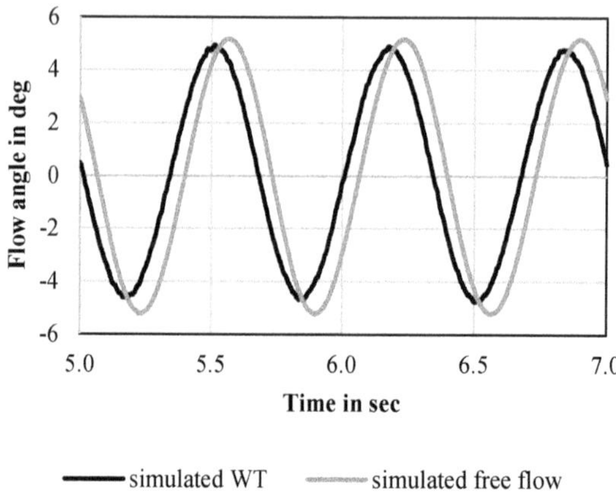

Figure 5.12: Flow angle time history 200 mm in front of the model at z = 150 mm; sine signal; f = 1.5 Hz, nominal A = 5°

Next, unsteady flow values will be evaluated in **Figure 5.13**. It shows the flow angle standard deviation for all signals at a nominal amplitude of 5°, as well as the percentage difference between both simulation environments.

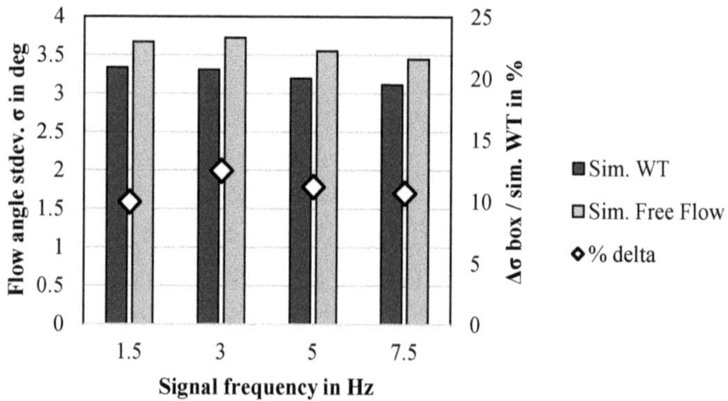

Figure 5.13: Flow angle standard deviation in front of the model at z = 150 mm; nominal amplitude A = 5°

The larger flow angle standard deviation in the free flow environment is visible for all signal frequencies. As established in section 4.3, the high yaw angles that occur during high standard deviation unsteady flow cause higher drag. Thus, the higher flow standard deviation is the main reason for the higher drag in the simulated free flow environment.

It can be argued that measuring flow angles 200 mm in front of the vehicle is not the same as measuring at the turntable center, which was used to determine the signal amplitude used for the free flow environment (see section 5.1.2). Thus, **Figure 5.14** shows the flow angle time history at the turntable center with the model present, at $z = 400$ mm and therefore placed in the unobstructed airflow above the model.

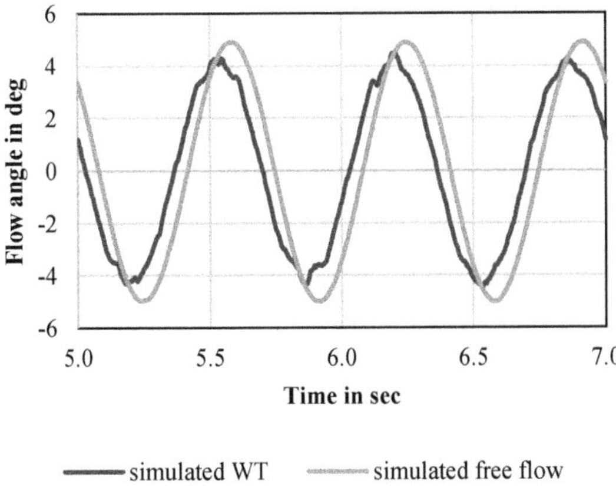

Figure 5.14:   Flow angle time history at turntable center, above DrivAer model; $z = 400$ mm; sine signal; $f = 1.5$ Hz, nominal $A = 5°$

The discrepancy between the amplitudes is still existent at turntable center in unobstructed flow. Additionally, the flow angle time history in the simulated wind tunnel no longer looks like an undisturbed sine signal.

To summarize, the analysis of the front end yielded the following information:

- Differences in the drag force development between simulated wind tunnel and free flow, as well as differences between different signal frequencies, can be explained by the nonzero and signal dependent static pressure distributions in the wind tunnel test section
- The signal amplitude in the free flow simulation was chosen so that it is equal to the amplitude in the empty test section of the simulated wind

tunnel. However, the presence of a vehicle changes the correlation between flap angle and flow angle in the wind tunnel. As a result, the vehicle no longer experiences the same flow amplitude in the simulated wind tunnel and free flow. This change in correlation between flap and flow angle is likely vehicle-dependent.

- At present, the previous point causes the flow angles experienced by the vehicle to be larger in free flow, leading to a higher drag in free flow.

5.4.2   Rear End

As shown previously, there is a big difference between the force development of the model in the simulated wind tunnel and in free flow at the vehicle rear end (**Figure 5.11**). Additionally, the rear end is a big contributor to differences between steady and unsteady flow, as well as differences between different signal frequencies (**Figure 5.9, Figure 5.10**).

As opposed to the general evaluation of corrected drag, here it is actually prudent to evaluate the uncorrected drag values because the flow field images and probes cannot undergo correction. Thereafter, it becomes possible to correlate uncorrected drag to flow parameters. **Figure 5.15** shows the uncorrected drag.

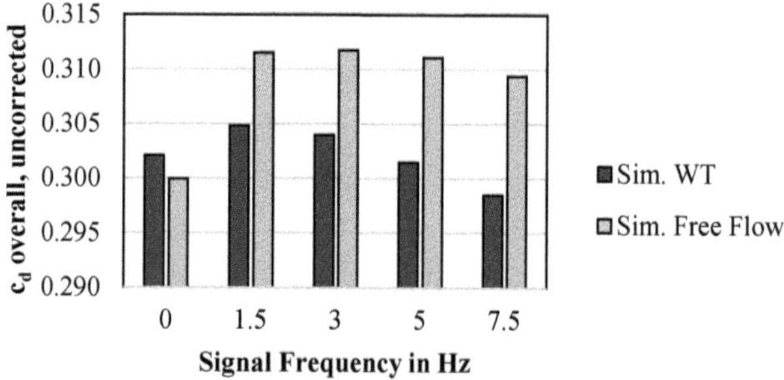

Figure 5.15:   Vehicle drag over frequency spectrum, uncorrected (nominal amplitude A = 5°; except steady-state flow for 0 Hz)

When unsteady flow is introduced at a low frequency, e.g. 1.5 Hz, vehicle drag increases in both the simulated wind tunnel and free flow environments compared to steady flow. The magnitude of this drag increase is much smaller in the simulated wind tunnel. There, the introduction of unsteady flow causes the pressure at the rear of the vehicle to rise (see e.g. **Figure 4.3**). This pressure increase counteracts the increase in drag due to yaw effects, explaining the smaller drag increase from steady to unsteady flow in the simulated wind tunnel. Further on, with increasing signal frequency (3–7.5 Hz), the drag values in both environments decrease again. **Figure 5.16** shows a top-down view of the model rear wake, which explains these phenomena. The blue region is the area with a total pressure less than zero, which is a good representation of wake size.

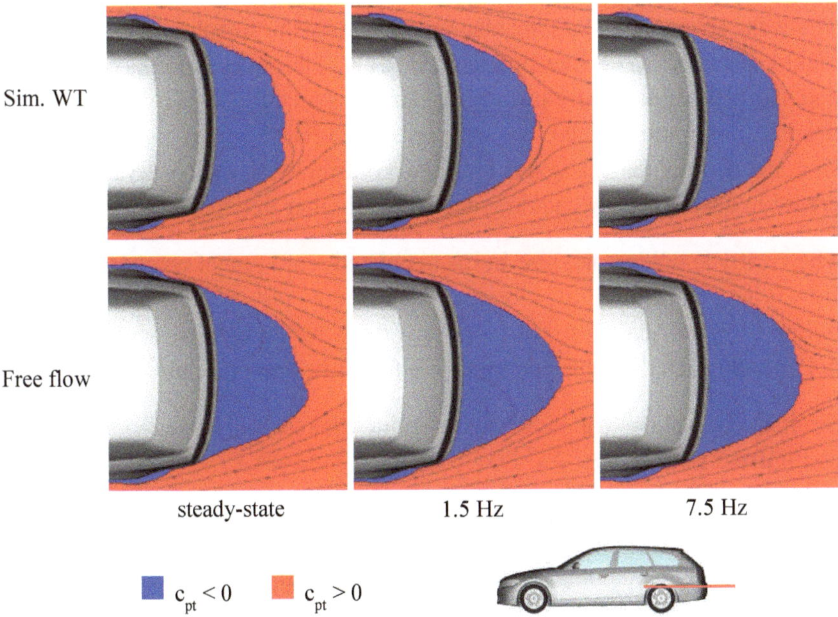

Figure 5.16: Rear wake, visualized via total pressure, top view, $z = 150$ mm (red line: image position)

Observation of the streamlines in the rear end wake reveals that the flow becomes more symmetrical and regular with increasing signal frequency (1.5 Hz to 7.5 Hz). This is especially evident in the free flow environment, where no interference effects are present. It is likely that this phenomenon plays a role in decreasing drag for higher frequencies. This reasoning is supported by

Bock's investigations into drag sources for squareback vehicles [77]. He concluded that a significant part of the aerodynamic drag originates from coherent structures generated in the wake. In order to break up these coherent structures, Bock utilized alternating, periodic flow injections on the left and right sides of the rear end wake [77]. Here, this is achieved inadvertently by the sine signal, which breaks up the rear end wake structures in a similar fashion. The observed symmetry and regularity of the flow in the wake at 7.5 Hz show that these higher frequencies actually decrease the magnitude of the aforementioned coherent structures. This is likely because higher frequencies do not allow the flow to settle in a pattern easily.

The wake shapes and sizes are also influenced by the signal frequency, as evident especially in the free flow environment. The wake is initially rounded and short. When a low-frequency sine signal is introduced (1.5 Hz) the wake becomes long and pointed, but with further increasing frequency (7.5 Hz) it shortens and rounds out again. This also contributes to the observed drag increase in unsteady flow and drag decrease at higher frequencies.

The clearly smaller wake size in the simulated wind tunnel compared to free flow is a direct result of the aforementioned empty test section pressure distribution. With increasing signal frequency, the wake in the simulated wind tunnel goes through the same changes as in free flow, but on a subtler scale.

Similar observations can be made observing the rear end wake from the side, as **Figure 5.17** shows.

As in the top view, the following points are also observed in the side view:

- Model in the simulated wind tunnel has a smaller wake due to the static pressure distribution (with high pressure near collector).
- Unsteady flow initially lengthens the wake at low frequency (1.5 Hz), but wake reverts in length with further increasing frequency (7.5 Hz).

However, the streamlines do not show any qualitative trends with increasing frequency. This supports the aforementioned hypothesis of the sine flow breaking up coherent structures inherent in steady flow: Since there is no external excitation in z-direction, the coherent structures visible in the side view here should be unaffected by the sine flow.

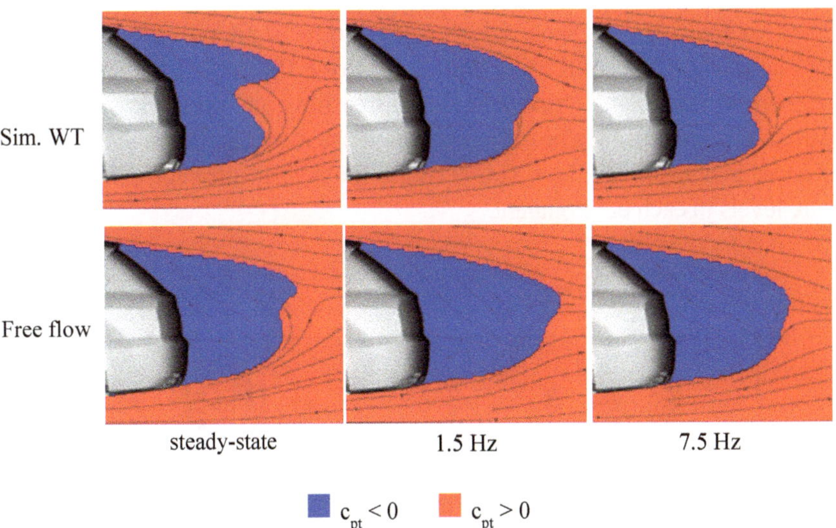

Figure 5.17:   Rear wake, visualized via total pressure, side view, $y = 0$

Up until this point, it has been established that while the wake size in the simulated wind tunnel and in free flow are quite different, this is mainly a result of the pressure distribution influence and can be accounted for using interference corrections. At the same time, the changes in wake shape and structure with changing signal frequency are very similar in the two environments. This makes the (simulated) wind tunnel a good predictor for the influence of unsteady flow on the vehicle wake. This hypothesis can further be confirmed by assessing flow probes at the rear end. **Figure 5.18** shows the probe locations chosen: One on the centerline at $z = 150$ mm (henceforth referred to as "rear middle" and two at the same height but at $y = \pm200$ mm (henceforth referred to as "rear left", "rear right" and collectively "rear side".

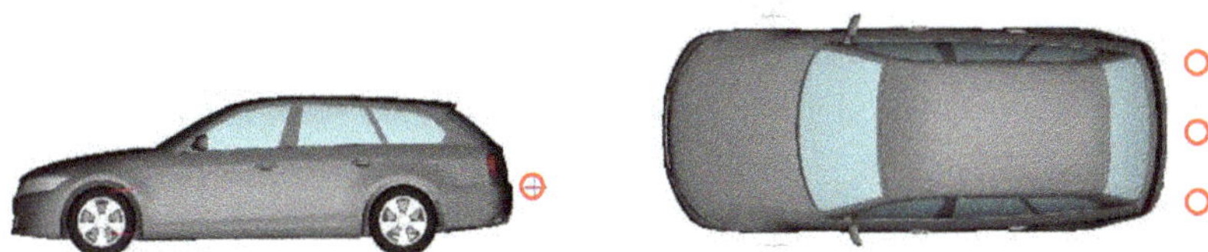

Figure 5.18:   Rear end CFD flow probe positions (50 mm behind rear bumper; at $y = 0$ and $y = \pm200$ mm, $z = 150$ mm)

The two environments show good correlation for static pressure at both the middle and the sides of the rear end, as **Figure 5.19** shows. In order to better compare the simulated wind tunnel and free flow, the pressure coefficients were normalized using a probe placed above the model at turntable center that sits within undisturbed flow. This accounts for slightly different absolute pressure levels in each environment.

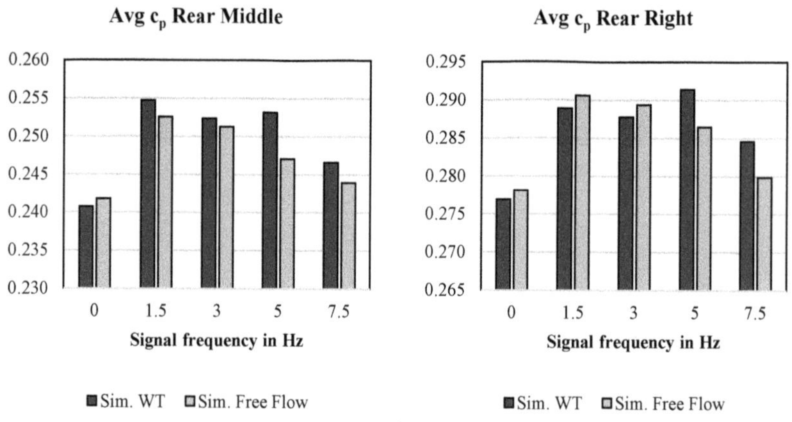

Figure 5.19:   Average static pressure rear middle (left image) and rear right (right image), normalized to free flow point above the model, at turntable center

Furthermore, the turbulent values also align between simulated wind tunnel and free flow, as exemplified by the turbulence intensity in y-direction shown in **Figure 5.20**.

Most importantly, the relationship between changing signal frequency and the respective flow values in **Figure 5.19** and **Figure 5.20** (pressure, Tu_y) remains similar between simulated wind tunnel and free flow, supporting the previous assertion of good correlation between the two. The same goes for the frequency profiles of flow angle and static pressure, of which the ones from the rear middle are shown in **Figure 5.21** and **Figure 5.22**.

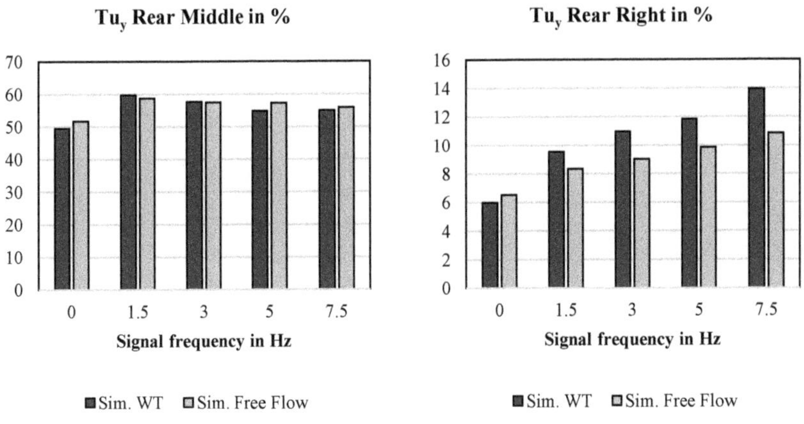

Figure 5.20: Turbulence intensity in y-direction at rear middle (left image) and rear right (right image)

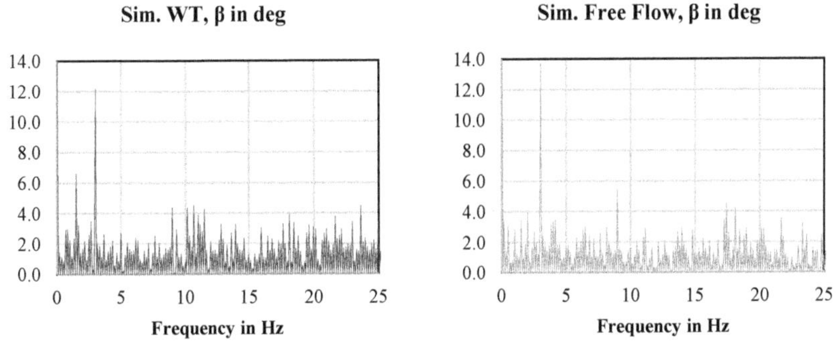

Figure 5.21: Flow angle FFT analysis at rear middle, sine signal at $f = 3$ Hz and nominal amplitude $A = 5°$, simulated wind tunnel (left) and free flow (right)

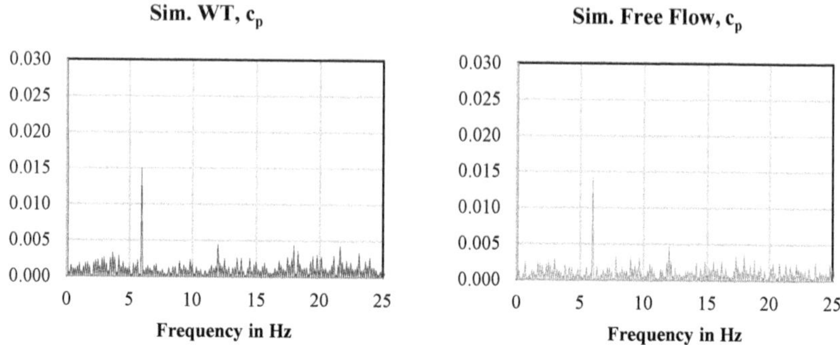

Figure 5.22:   Static pressure FFT analysis at rear middle, sine signal at f = 3 Hz and nominal amplitude A = 5°, simulated wind tunnel (left) and free flow (right)

The frequency profiles show peaks at signal frequency (3 Hz) for flow angle and double signal frequency (6 Hz) for pressure, which is to be expected since pressure is a scalar value that yields roughly the same value for both directions of the sine flow. Neither flow environments show significant other frequency peaks.

To summarize, the analysis of the rear end yielded the following information:

- Similar to the front end, differences in the drag force development between simulated wind tunnel and free flow can be explained by the signal-dependent static pressure distribution in the wind tunnel test section.
- Changing signal frequencies change the wake shape at the rear of the vehicle, which is directly correlated to resultant drag.
- After introducing low-frequency unsteady flow, the wake increases in size and assumes a smoother shape, but gradually returns towards the original size and shape with increasing signal frequency. Consequently, drag follows a similar pattern, increasing from steady to low frequency unsteady flow but decreasing with further increasing frequency.
- While absolute drag values and wake sizes are very different between simulated wind tunnel and free flow due to gradient influence, the two environments behave very similarly otherwise, as evidenced by comparison of turbulent parameters obtained from flow probes. Currently, it is not possible to correct for static pressure gradients when analyzing 3D

flow fields, since contemporary correction methods only use superposition methods along a 1D line to calculate overall forces [15] [16] and therefore are ill suited for 3D corrections.

5.4.3   Front Wheels

The previous sections detailing the front and rear ends suggest that differences between wind tunnel and free flow are either minor or can be overcome using correction methods. However, for the last relevant section investigated, the front wheels, this is not the case.

The velocity profile from the side shown in **Figure 5.23** also shows the differences between the two simulation environments.

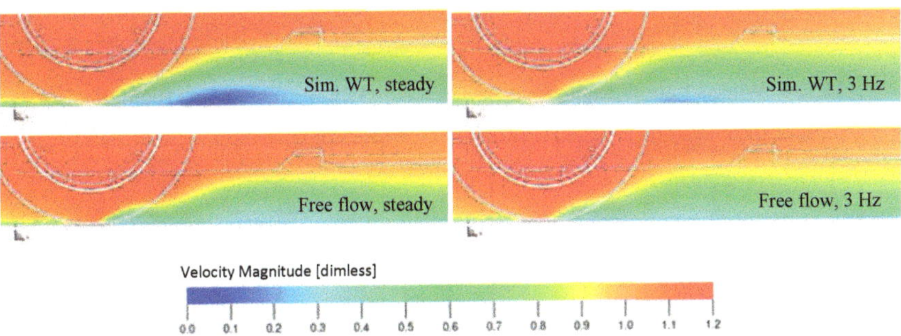

Figure 5.23:   Velocity side profile near the front wheels, at $y = 250$ mm, flow moving left to right

There is a clear difference between the velocity profile for steady and unsteady flow in the simulated wind tunnel, while the difference in free flow is minuscule. This shows that the flow regime near the floor and behind the wheels to be very different between the two simulation environments.

Like for the front and rear ends previously, further quantitative analyses can be made using flow probes, whose locations are shown in **Figure 5.24**: Two on each side of the model at $y = \pm 250$ mm ("left" and "right"), of which one is located at $z = 50$ mm ("bottom") and one at $z = 150$ mm ("top").

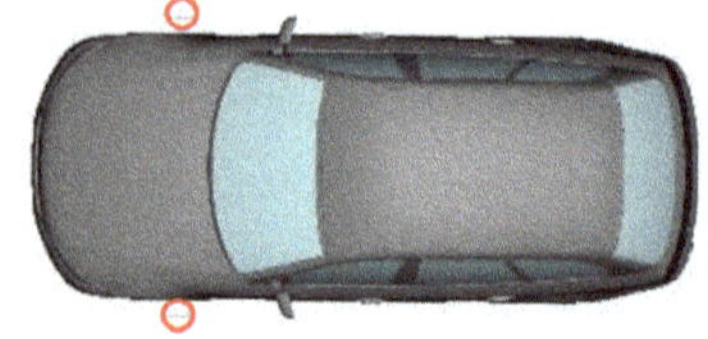

Figure 5.24:   Front wheel CFD flow probe positions (300 mm in front of turntable center, at $y = \pm 250$ mm and $z = 50$ mm and $z = 150$ mm)

The difference between simulated wind tunnel and free flow near the floor is best represented by the turbulent length scale, shown in **Figure 5.25** in y-direction for both the top and bottom probes on the right side. The length scale represents the size of large eddies in the local turbulent flow extending in y-direction.

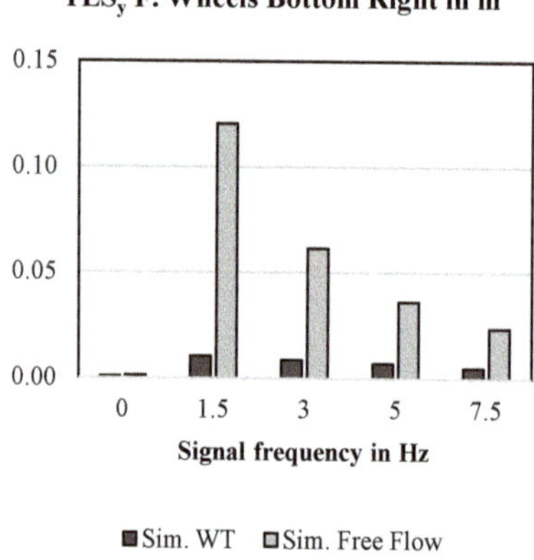

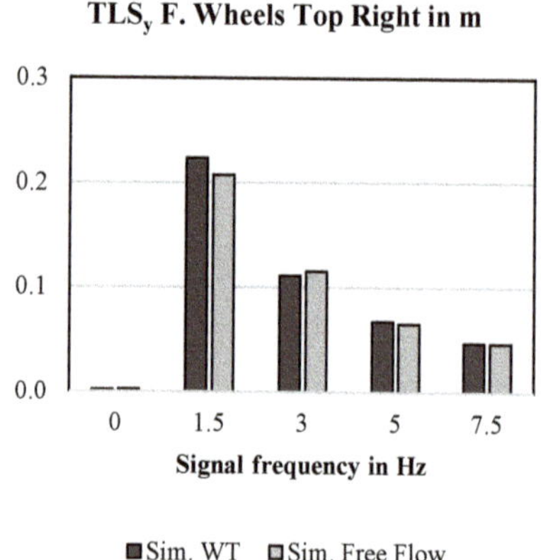

Figure 5.25:   Turbulent length scale in y-direction, front wheels, $z = 50$ mm (left) and $z = 150$ mm (right)

The eddies in free flow near the ground ("bottom") are much larger than those in the simulated wind tunnel, while the same effect is not present farther off the ground ("top").

Frequency profiles taken at z = 50 mm further show the different near ground behavior in the simulated wind tunnel and free flow environments. **Figure 5.26** shows the flow angle FFT in both environments.

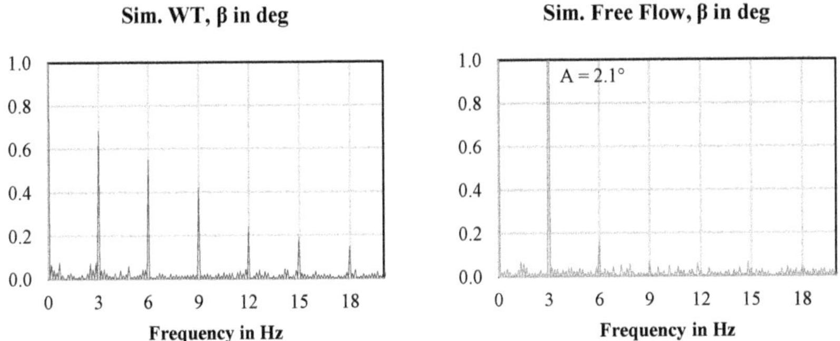

Figure 5.26: Flow angle FFT analysis, front wheels, sine signal at f = 3 Hz and nominal amplitude A = 5°, z = 50 mm, simulated wind tunnel (left) and free flow (right)

Like in **Figure 5.25** previously, the difference between the environments is clearly visible. The free flow environment shows a high, single peak at signal frequency, while the simulated wind tunnel shows multiple peaks at multiples of the signal frequency. At z = 150 mm (not shown here), however, both FFT results are the same, revealing the differences in **Figure 5.26** to also be caused by floor interactions.

Similar to the flow analysis at the rear end, flow probes can also analyze the relationship between changing signal frequency and turbulent flow values, comparing them in both environments. **Figure 5.27** shows the pressure fluctuation, while **Figure 5.28** shows the turbulence intensity in y-direction, at both probe heights.

It is evident that the effect of increasing signal frequency on pressure fluctuation and turbulence intensity is the opposite in the simulated wind tunnel compared to free flow. Generally, the aforementioned unsteady values decrease with increasing frequency in free flow, but increase with increasing frequency in the simulated wind tunnel. This effect is present at both probe heights.

Therefore, it is not contingent on floor interactions like some of the previous observations are.

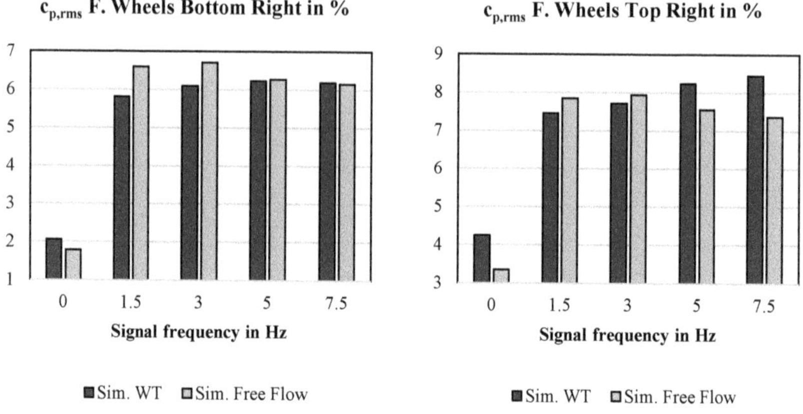

Figure 5.27:   Pressure fluctuation, front wheels, right side probes, $z = 50$ mm (left) and $z = 150$ mm (right)

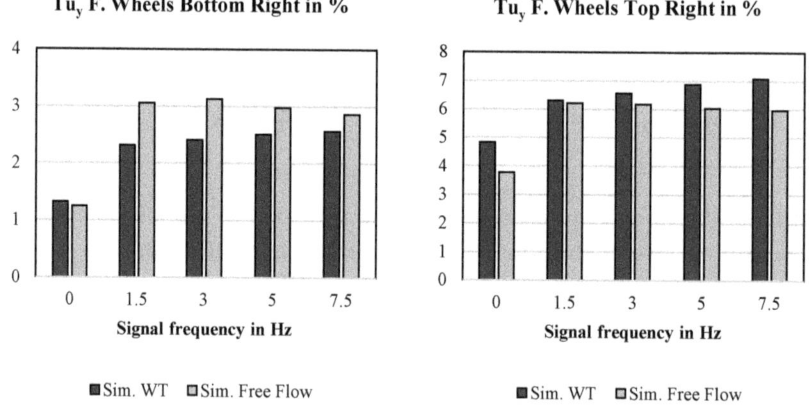

Figure 5.28:   Turbulent intensity in y-direction, front wheels, right side probes, $z = 50$ mm (left) and $z = 150$ mm (right)

The lower fluctuations and turbulence in free flow with increasing signal frequency can be explained by the shorter periods at higher frequencies. The airflow at the wheels has less time to settle in one particular direction before the flow reverses direction. This lowers overall turbulence compared to lower frequency signals of the same amplitude. This reduction in turbulence with higher frequency is similar to the rear end phenomenon discussed in section 5.4.2. There, the wake size and many of the unsteady values decrease with increased signal frequency as well.

Consequently, the opposite development present in the simulated wind tunnel must be rooted in interactions with wind tunnel specific phenomena. The most likely of these is the shear layer present at the sides of the wind tunnel jet, since they are closer to the front wheel areas but farther away from the rear end wake. This would explain why the aforementioned discrepancies do not occur at the rear end separation zone. However, the exact mechanisms of the shear layer interaction cannot be discerned from the results up until this point and will be discussed in more detail later on.

To summarize, the analysis of the front wheel area yielded the following information:

- Flow field and flow probe analyses show a substantial difference between the two environments near the floor (qualitative flow fields and turbulent values) that correlate to the above point.
- Flow values shows opposing behavior in the two environments when exposed to increasing signal frequency. Pressure fluctuations and turbulence intensity increase with increasing signal frequency in the simulated wind tunnel, while decreasing with increasing signal frequency in free flow. This is independent of probe proximity to the ground.
- The reason for the substantial differences between simulated wind tunnel and free flow likely lies in interactions of the flow around the vehicle with the floor (boundary layer) and in effects occurring in the shear layers at the sides of the wind tunnel jet. The specifics will be discussed in the next section.

5.5   Cause of Discrepancies between Wind Tunnel and Free Flow

So far, the following has been established concerning correlation between the simulated wind tunnel and free flow environments:

- There are some differences at the front end between wind tunnel and free flow. These either do not strongly influence deltas between different configurations/signals or can also be corrected using interference corrections.
- The rear end behaves similarly for both simulation environments. Differences in overall drag and wake pressure levels can be corrected using interference corrections.
- The front wheel area, and by extension areas at the sides of the vehicle, react differently to both the introduction of unsteady flow and to changes in unsteady flow frequency.

In order to better understand the observations made at the front wheel area, additional empty test section simulations were undertaken at a sine frequency of 1.5 Hz, in both the simulated wind tunnel and free flow environments. Flow probes were placed around the turntable center ($x = 0$) to compare the basic flow situations without vehicle interactions. The additional simulations have revealed three main causes of the differences between wind tunnel and free flow:

- Loss of velocity magnitude away from jet centerline
- Interaction of swing airfoil with wind tunnel floor
- Interaction with coherent structures in shear layer

5.5.1   Loss of Velocity Magnitude away from Jet Centerline

In a free flow setup, the incident flow's velocity magnitude and total pressure will be the same everywhere before the vehicle is introduced, no matter its flow angle. However, this is not the case in an open jet wind tunnel, simulated or real, because the jet's size is limited. The wind tunnel flow consists of a core jet with the desired freestream velocity, the plenum area outside of the jet at ambient pressure and a velocity of zero and a shear layer with variable velocity and pressure that separates the two. **Figure 5.29** shows this for the simulated wind tunnel.

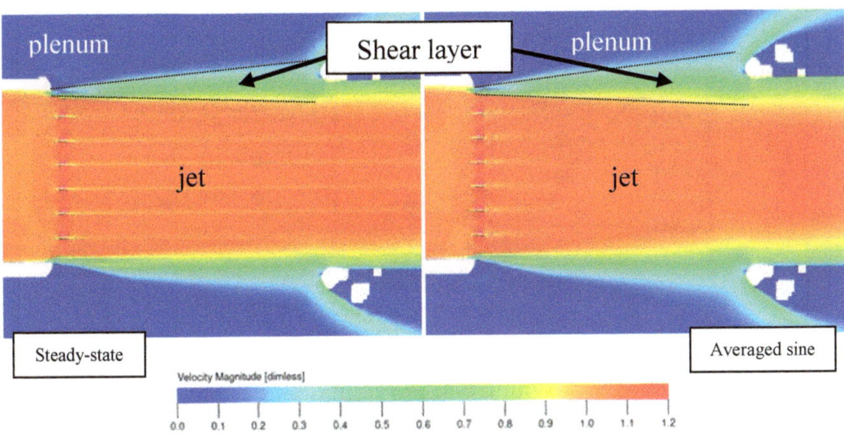

Figure 5.29:　Velocity magnitude in the empty test section, simulated wind tunnel, at z = 250 mm. Steady-state flow (left) and averaged sine flow at f = 3 Hz and nominal amplitude A = 5° (right)

When FKFS swing introduces a sideward component in the flow, the core jet and shear layers will move side to side accordingly. On average, this results in an increased shear layer size where the velocity magnitude is significantly lower than the desired freestream velocity. Consequently, the inner edge of the shear layer will be closer to the center of the test section. Both effects are evident when comparing the two pictures in **Figure 5.29**.

However, when analyzing the probe data from the empty test section, it is evident that this loss of velocity magnitude does not only occur in the enlarged shear layer, but also within the core jet when swing is active. **Figure 5.30** shows that while in free flow the velocity magnitude stays constant no matter the y-position of the probe, in the wind tunnel it rapidly decreases even at only 200 mm from the centerline. At y = 800 mm, the flow probe is already within the shear layer, which explains the much larger drop to about 0.65 (outside the scale of the diagram).

When observing the velocity magnitude time history for the simulated wind tunnel in **Figure 5.31** (y = 200 mm), it is clear that this loss of velocity directly results from the limited lateral jet size in the wind tunnel.

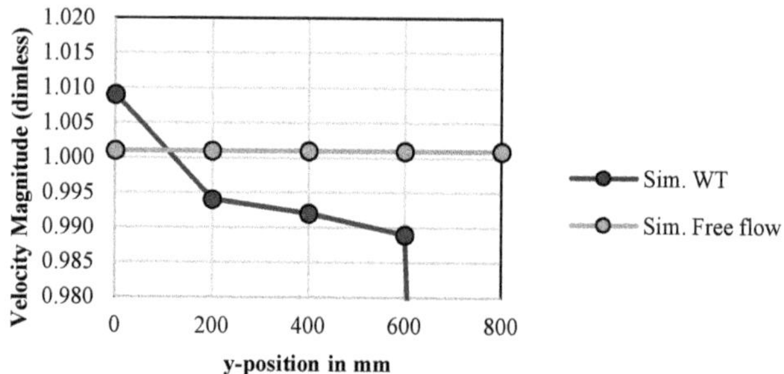

Figure 5.30:   Velocity magnitude in the empty test section, at x = 0 (turntable center), z = 250 mm. y = 800 mm: maximum extension of WT nozzle

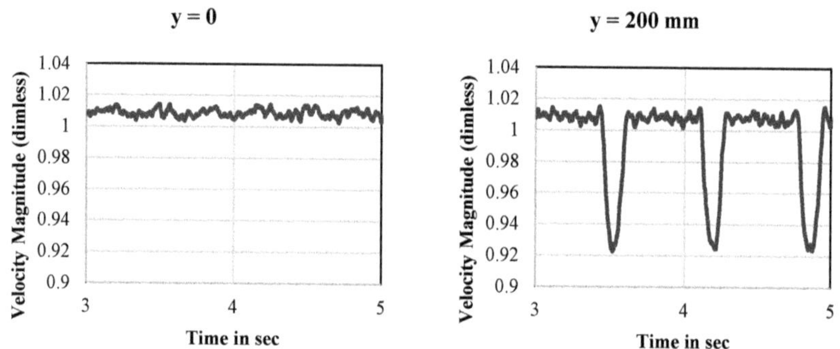

Figure 5.31:   Velocity magnitude time history, simulated WT, jet centerline (left) and 200 mm off centerline (right)

While the velocity magnitude on the centerline stays constant over time, off the centerline it shows sharp dips to about 0.92 aligning with the signal frequency of 1.5 Hz. This shows that even at y = 200 mm, which still lies within most 25 % scale models' widths, the wind tunnel jet cannot retain a constant nominal velocity magnitude.

As this loss of velocity only occurs off the jet centerline and gets progressively stronger farther off the centerline, it serves as one explanation why the resultant flow at the sides of the vehicle is different between wind tunnel and free flow.

5.5.2   Interaction of swing Airfoil with Wind Tunnel Floor

While general differences between wind tunnel and free flow can be attributed to the aforementioned velocity loss, the specific discrepancies occurring only near the ground can be explained by the interaction of the swing airfoils with the wind tunnel floor. The physical airfoils have a finite vertical extension, thereby creating a 3 mm gap between their bottom edges and the ground. This is not present in free flow, where the unsteady flow is created by time-resolved boundary conditions at the inlet.

This gap between the airfoil edge and the ground causes the achieved flow angle near the ground to be much lower than further up, as shown in **Figure 5.32**.

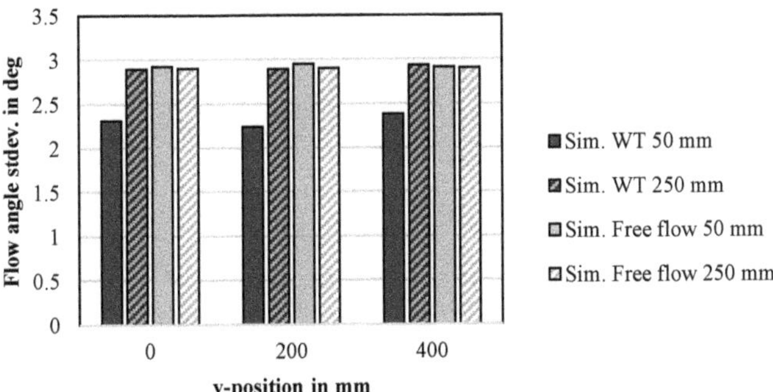

Figure 5.32:   Flow angle standard deviation at different heights, $x = 0$ and $y = 0$, empty test section, sine signal $f = 1.5$ Hz, nominal amplitude $A = 5°$

The flow angle standard deviation is the same in free flow for both probe heights, as well as for the simulated wind tunnel at $z = 250$ mm. Only at z = 50 mm in the wind tunnel is it substantially lower. This difference is present for all y-positions.

As mentioned, this discrepancy between wind tunnel and free flow is likely responsible for the undesirable flow behavior occurring near the ground discussed in section 5.4.3 and shown in **Figure 5.23**, **Figure 5.25** and **Figure 5.26**.

5.5.3   Interaction with Coherent Structures in Shear Layer

What remains is a more specific explanation of the differences between wind tunnel and free flow at the front wheel region when it comes to reactions to signal frequency changes (see **Figure 5.27**, **Figure 5.28**). Section 5.4.3 has postulated that this phenomenon is connected to the shear layer in the wind tunnel. Said shear layer has been a topic of research in the past, albeit not involving dynamic turbulence generation devices. Schönleber [78] has extensively discussed the shear layer in his dissertation, showing different coherent structures occurring in the wind tunnel jet that worsen measurement accuracy and reproducibility if not considered. The most relevant phenomenon he described in the context of the present research is the "flutter" flow structure, shown in **Figure 5.33**.

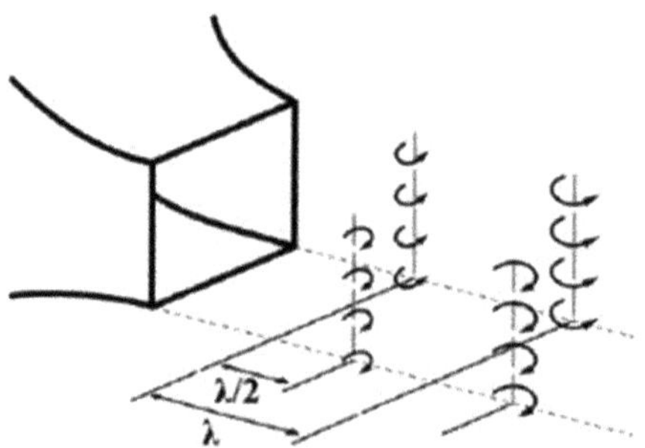

Figure 5.33:   "Flutter" flow structure with alternating vortices in left and right shear layers [78]

The shear layer contains high flow velocity on the jet side and low flow velocity on the plenum side (see **Figure 5.29**), creating pressure deltas that cause the fluid to travel laterally and rolling up into vortices. Depending on wind tunnel geometry and air speed, the vortices on both sides can be in phase or offset by half a phase, the latter characterizing the "flutter" phenomenon.

Shear layer vortices can be avoided using measures at the nozzle exit such as delta wings [78]. They have been installed in the MWK at the time of the current research and also modeled in the simulated wind tunnel. However, by forcibly adding a sine flow into the jet using the swing system, it is likely that "flutter" is reintroduced into the flow. In particular, many fluid values in sine flow are in fact offset by half a phase between left and right, such as y-velocity and flow angle.

The evidence of "flutter" present in the simulated wind tunnel during sine flow can be seen in the z-vorticity time history shown in **Figure 5.34**.

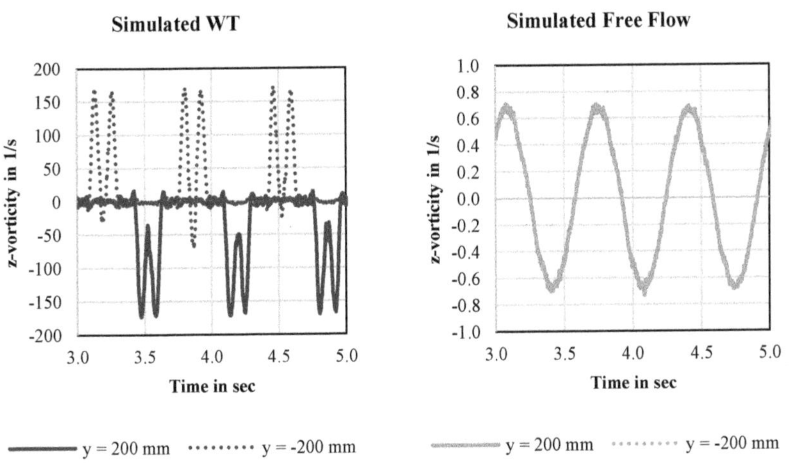

Figure 5.34: z-vorticity time history at turntable center, f = 1.5 Hz, nominal amplitude A = 5°, y = ±200 mm, z = 250 mm

The data was taken at z = 250 mm, far enough off the ground that ground interactions are not a concern. Two observations can be made here:

- z-vorticity peaks are much larger in the simulated wind tunnel (about two orders of magnitude).

- The left and right sides in the simulated wind tunnel are offset by half a phase and their peaks have opposing signs, while in free flow the left and right sides are identical.

Both these observations support the presence of "flutter" in the shear layer while under generated sine flow:

- The vorticity peaks in free flow only result from the changes in flow direction, which only occurs between ±5°, leading to low vorticity. In the wind tunnel, they result from the much stronger shear layer vortices.
- In free flow, the vorticity direction and phase are the same for both sides, because the flow angle time history is the same for both sides. In the wind tunnel, the shear layer vortices on both sides have opposite signs and are offset by half a phase (see **Figure 5.33**).

Furthermore, the aforementioned observations and conclusions are valid for all lateral positions, as **Figure 5.35** exemplifies using the same vorticity time history at $y = \pm 400$ mm.

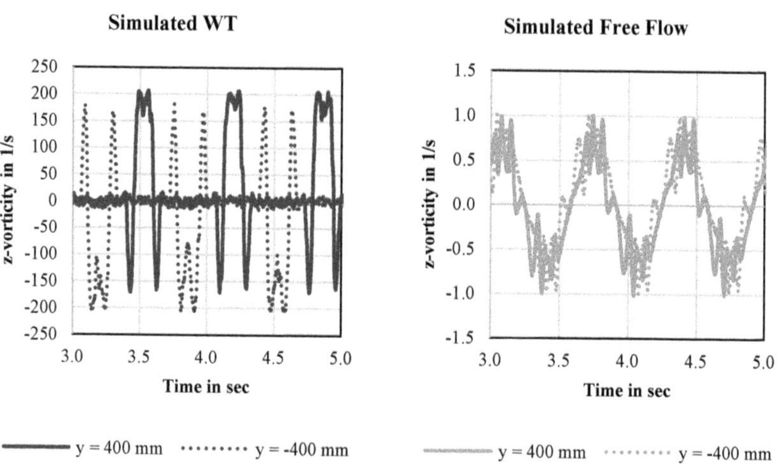

Figure 5.35:   z-vorticity time history at turntable center, $f = 1.5$ Hz, nominal amplitude $A = 5°$, $y = \pm 400$ mm, $z = 250$ mm

The "flutter" phenomenon therefore explains the increase in turbulent values near the front wheels in the simulated wind tunnel with increasing signal frequency. At higher frequencies, the period and wavelength of the "flutter" structure in the shear layer become shorter. Consequently, more vortices act on the flow field around the model at the same time, thereby increasing turbulence near the sides of the vehicle. The front and rear of the vehicle are not affected due to their higher physical distance from the shear layer.

5.6   Operating Conditions of the Gust Generation System

Previously, section 5.4 has established that while overall drag is comparable between wind tunnel and free flow, the concrete flow regimes at the sides of the vehicle are not. The reasons for this discrepancy have been outlined in section 5.5, with the biggest factor being the proximity of the vehicle sides to the sides of the limited size jet. This causes the lowered velocity (section 5.5.1) and the shear layer vortices (section 5.5.3) to strongly interact with the flow at the vehicle sides.

For now, it has to be accepted that this discrepancy between wind tunnel and free flow exists. However, knowing that the proximity to the edges of the jet is the root cause, it becomes possible to define operational conditions for the gust generation system. Within these boundaries, differences to free flow can be held to a minimum.

The limits of the swing system in the Model Scale Wind Tunnel are amplitudes up to 10° and frequencies up to 12 Hz (Section 4.1.1). High frequencies themselves do not present a problem: Issues only arise when high frequencies are combined with high amplitudes (see pressure distributions in section 4.2.1) or when discussing how changing frequencies relate to other results (front wheels, section 5.4.3). High amplitudes, however, directly influence the shape of the jet, specifically narrowing the core jet where flow velocity is high (see **Figure 5.29**). Therefore, it is prudent to research whether nominal sine amplitudes lower than 5° produce better correlation between simulated wind tunnel and free flow. The results are shown in **Figure 5.36**.

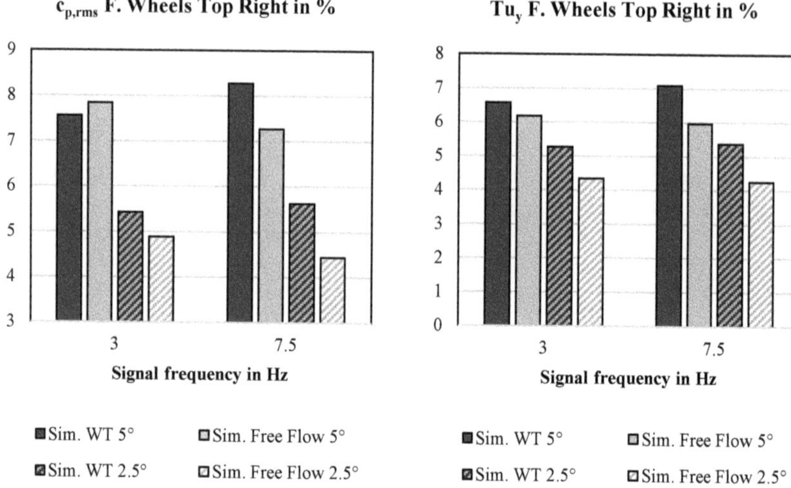

Figure 5.36: Pressure fluctuation (left) and turbulence intensity in y-direction (right), for different signal amplitudes, front wheel probe, 300 mm in front of turntable center, y = 250 mm, z = 150 mm

While for both 5° and 2.5° nominal amplitude the turbulent values increase from 3 to 7.5 Hz in the simulated wind tunnel and decrease in free flow, the discrepancy is not as large for the smaller amplitude. Table 5.2 below shows the exact numbers for the front wheel probe shown in **Figure 5.36**.

Table 5.2: Percentage point changes when raising sine frequency from 3 to 7.5 Hz, for different signal amplitudes

Nominal amplitude A in °	Simulation environment	$c_{p,rms}$ change from 3 Hz to 7.5 Hz	Tu$_y$ change from 3 Hz to 7.5 Hz
5.0	Sim. WT	+ 0.7 %	+ 0.5 %
	Free flow	− 0.5 %	− 0.2 %
2.5	Sim. WT	+ 0.2 %	+ 0.1 %
	Free flow	− 0.4 %	− 0.1 %

The numbers show that the difference between 3 and 7.5 Hz is approximately halved for the lower amplitude. Similar results are achieved for other probes and turbulence values at the model's sides.

At a typical highway speed of 120 km/h, cars can expect an incident flow with roughly 2.2° standard deviation [70]. This is achieved by a sine amplitude of 3.1°. Therefore, the negative effects the shear layer has on correlation between wind tunnel and on-road flow will not present a substantial issue for measurements at typical speeds and incident flow angles. Flow simulating natural turbulence will also in all likelihood not cause issues, since those are generally lower than on-road turbulence involving traffic. It is, however, prudent to caution future users of the swing system against operating at standard deviations significantly higher than those 2.2°.

5.7   Summary

At the start of this chapter, the following questions were asked:

1.   How accurate are measurements with FKFS swing compared to blockage-free on-road drag values? How big the differences and what are their causes?
2.   How do different signal frequencies result in different unsteady drag in vehicles with high drag sensitivity?

Concerning the comparison between wind tunnel and free flow, CFD investigations have uncovered the following:

- Absolute (corrected) drag is quite different between the two, but relative to each other the correlation is good.
- Differences in force development between wind tunnel and free flow are largely caused by different static pressure distributions in the wind tunnel. This can be remedied using interference corrections. Non-pressure distribution reasons for differences follow below.
- The relation between flap angle and flow angle in the wind tunnel is dependent on the vehicle geometry. In order to correlate wind tunnel to free flow, the applied flow signal in free flow also needs to adapt for different vehicles.

- Wind tunnel and free flow show similar flow fields and turbulence values at the vehicle's rear end.

- Near-ground flow and flow at the vehicle sides are markedly different between wind tunnel and free flow; especially when it comes to changes caused by changes to signal frequency.

- The reason for the difference near the ground is the interaction of the airfoil edge with the ground, since a gap exists between airfoil and ground.

- The difference at the vehicle sides are caused by two wind tunnel interactions: The loss of velocity magnitude away from the jet centerline; the interactions of the flow with the coherent structures / vortices in the shear layer.

- The differences between wind tunnel and free flow can be reduced greatly by limiting signal amplitude. Common on-road flow signals will thus have good correlation between wind tunnel and free flow.

Concerning the mechanisms on how frequency changes affect vehicle drag, the following results were found:

- Different signal frequencies create different wake shapes, especially in the rear of the vehicle.

- The wake shapes correlate with drag: larger wakes result in higher drag.

- Initially in steady-state flow, the wake shape is fairly irregular and bulbous. At low-frequency unsteady flow, the wake increases greatly in size and becomes more pointed. With further increasing frequency, however, the wake gradually returns to its steady-state shape and size.

- The previous observation mirrors drag changes: A big increase in drag from steady-state to low-frequency unsteady flow, then drag decrease with further increasing frequency

- The initial large wake and high drag at low-frequency unsteady flow stems mainly from the sideward, yawed flow introduced. The subsequently decreasing wake size and drag stem from suppression of natural coherent structures in the vehicle wake due to the forced excitation, and from lessened turbulent energy levels / eddy sizes with increasing frequency.

6 Summary and Conclusion

Today, economic and ecologic concerns drive aerodynamic development in production vehicles. While drag optimization is carried out in steady-state environments, it does in fact not represent actual on-road drag well, as the on-road flow state is characterized by unsteady turbulence. In order to evaluate realistic vehicle drag, gust generation systems in wind tunnels have been used in the past, as well as CFD simulations in a blockage-free environment. Generally, those works have utilized a handful of unsteady flow signals, usually derived from on-road measurements, and fairly limited vehicle geometry variations.

The goals of this work were:

- Expand on both the unsteady signals used and the vehicle geometry variations in order to create a better understanding of the influence of unsteady flow on passenger car drag.
- Learn more about the effects of an active gust generation system in the wind tunnel.

To this end, the presented research is the first one that includes a large (10+) number of flow signals, parametrized using frequency and flow standard deviation, as well as a large number of geometry variations in the measured vehicle models. Furthermore, some of the measurements were recreated in CFD, using both a blockage-free as well as a simulated wind tunnel environment that replicates the actual model scale wind tunnel used in the experimental measurements.

The experiments and simulations undertaken have resulted in a number of findings:

- The active generation system creates a static pressure distribution in the test section that is both different from the steady-state static pressure gradient and unique to the applied unsteady incident flow signal. Consequently, to obtain valid drag measurements and comparisons, interference corrections are always necessary. Contemporary open-jet wind tunnel interference correction has been modified for usage with unsteady measurements and subsequently validated.

© The Author(s), under exclusive license to
Springer Fachmedien Wiesbaden GmbH, part of Springer Nature 2026
X. Fei, *The Impact of Unsteady Flow on Drag Measurements in Automotive Wind Tunnels*, Wissenschaftliche Reihe Fahrzeugtechnik Universität Stuttgart,
https://doi.org/10.1007/978-3-658-51766-3_6

- Unsteady incident flow influences vehicle drag in two ways: Yaw angle magnitude and frequency spectrum. The influence of yaw angle magnitude broadly conforms to the drag behavior of vehicles under steady-state yaw flow, like during measurements with rotated turntables. Higher magnitudes generally increase drag. The frequency effect, on the other hand, cannot be reproduced using steady-state methods. At the same average yaw angle, two signals with different frequency profiles can still result in differing drag, depending on vehicle geometry.

- In order to assess the frequency effects on drag, two vehicle quantifiers were introduced: Drag Sensitivity (S_D) and Average Delta to Quasi-Steady ($\Delta_{QS,avg}$). High S_D means that the vehicle experiences big changes in drag when the signal frequency profile is changed, while high $\Delta_{QS,avg}$ means that the vehicle's unsteady drag cannot be accurately estimated using steady-state methods. A number of geometric traits were determined to affect Drag Sensitivity and Delta to Quasi-Steady. Among those are the rear end shape (squareback or notchback), the vehicle separation area, specific characteristics of the rear end geometry, and the presence and design of the wheel areas.

- The drag reactions to unsteady flow related to high S_D and $\Delta_{QS,avg}$ are caused by the unsteady flow's influence on vehicle wakes. Among these wakes are those present at the rear end, wheels and mirrors. Specifically, low-frequency unsteady flow has the biggest effect on wake shapes and on drag due to high turbulent energy, while higher frequencies tend to return the wake shape towards the original steady-state shape.

- By comparing CFD results in a simulated wind tunnel environment to results in a blockage-free box environment, it was determined that the FKFS swing gust generation system is generally able to accurately measure unsteady averaged drag. This is confirmed by well matching flow fields and turbulent values at the rear end of the vehicle.

- At high amplitudes for sine signals, coherent vortex structures in the shear layer change the flow field at the sides of the vehicle compared to a blockage-free box simulation. However, this phenomenon is amplified by the sine signals chosen for the investigations, which are not representative for on-road driving. Realistic on-road signals contain many discrete frequencies instead of just one and also possess lower signal amplitudes, thereby lessening the periodic interactions mentioned.

From the findings, the following conclusions can be inferred for the testing and optimization of automobile aerodynamics:

- The drag of any vehicle under a specific unsteady flow situation must be measured individually. It cannot always be inferred from measurements made on a different vehicle, under a different unsteady flow signal, or using steady-state approximations such as quasi-steady drag.
- When using a gust generation system, the signal amplitude must be limited in order to achieve a flow field comparable to on-road flow. For the swing system, specifically, using signals emulating natural turbulence or on-road turbulence with traffic will be unproblematic.

In order to further the research undertaken in this work and to improve procedures for unsteady aerodynamic testing, further steps that can be taken include:

- Investigations on how different vehicles in the test section influence the transfer function from swing flap angle to flow angle for different unsteady flow signals
- Improvements on correlation between the flow fields of CFD and wind tunnel; not only with respect to streamwise static pressure distribution and averaged drag, but also to flow parameters in the whole test section. Flow probe measurements in the test section with different flow signals and subsequent adjustments of CFD setup to reflect the measurement results
- In-depth investigations on vehicle flow fields under unsteady flow, e.g. using surface pressure probes, wake probes, or Particle Image Velocimetry
- Quantification of inaccuracies in the wind tunnel compared to on-road flow, dependent on flow signal and vehicle geometry
- Creation of guidelines for industrial aerodynamic testing concerning the limits of gust generation systems and their margin of error

The listed findings contribute a great amount to understanding both the effects of unsteady flow on passenger car drag, as well as the intricacies of operating an active gust generation system in an automobile wind tunnel for aerodynamic testing. This work can furthermore serve as a foundation for both future research on these topics, as well as a basis to develop a validated and robust process for aerodynamic development of production cars in realistic, unsteady flow environments.

Bibliography

[1] T. Woll, Einfluss der Aerodynamik auf den Verbrauch heutiger und zu-künftiger Fahrzeugkonzepte, München, 2001.

[2] *Commission Regulation (EU) 2017/1151 of 1 June 2017 supplementing Regulation (EC) No 715/2007 of the European Parliament and of the Council on type-approval of motor vehicles,* 2017.

[3] J. Wiedemann, „Leichtbau bei Elektrofahrzeugen," *ATZ Automobil-technische Zeitschrift 111,* pp. 462-463, June 2009, doi:10.1007/BF03222084.

[4] C. Jessing, Charakterisierung instationärer Anströmungen und Analyse ihrer Einflüsse auf die Fahrzeugaerodynamik, Doctoral Dissertation, Universität Stuttgart, Springer Fachmedien Wiesbaden GmbH, 2021, doi:10.1007/978-3-658-34847-2.

[5] W.-H. Hucho, "Kapitel 1 - Einführung," in *Hucho-Aerodynamik des Automobils: Strömungsmechanik, Wärmetechnik, Fahrdynamik, Komfort,* Springer-Verlag, 2013.

[6] A. Dillmann, "Kapitel 2 - Physikalische Grundlagen der Aerodynamik," in *Hucho-Aerodynamik des Automobils: Strömungsmechanik, Wärme-technik, Fahrdynamik, Komfort,* Springer-Verlag, 2013.

[7] S. Wordley and J. Saunders, "On-road Turbulence," *SAE Int. J. Passeng. Cars - Mech. Syst.,* vol. 1, no. 1, pp. 341-360, 2009, doi:10.4271/2008-01-0475.

[8] P. O'Neill, D. Nicolaides, D. Honnery and J. Soria, "Autocorrelation Functions and the Determination of Integral Length with Reference to Experimental and Numerical Data," in *15th Australasian fluid mecha-nics conference,* Sydney, NSW, Australia, 2004.

[9] G. Taylor, "Statistical theory of turbulence," *Proc. R. Soc. A.,* vol. 151, no. 873, pp. 421-444, 1935, doi:10.1098/rspa.1935.0158.

© The Editor(s) (if applicable) and The Author(s), under exclusive license to Springer Fachmedien Wiesbaden GmbH, part of Springer Nature 2026
X. Fei, *The Impact of Unsteady Flow on Drag Measurements in Automotive Wind Tunnels,* Wissenschaftliche Reihe Fahrzeugtechnik Universität Stuttgart,
https://doi.org/10.1007/978-3-658-51766-3

[10] S. W. Smith, "Chapter 8: The Discrete Fourier Transform," in *The Scientist and Engineer's Guide to Digital Signal Processing*, San Diego, Calif., California Technical Publishing, 1999.

[11] R. Blumrich, E. Mercker, A. Michelbach, J.-D. Vagt, N. Widdecke und J. Wiedemann, „Kapitel 13 - Windkanäle und Messtechnik," in *Hucho-Aerodynamik des Automobils: Strömungsmechanik, Wärmetechnik, Fahrdynamik, Komfort,*, Springer-Verlag, 2013.

[12] R. Blumrich, N. Widdecke, J. Wiedemann, A. Michelbach, F. Wittmeier and O. Beland, "New FKFS Technology at the Full-Scale Aeroacoustic Wind Tunnel of University of Stuttgart," *SAE Int. J. Passeng. Cars - Mech. Syst.*, vol. 8, no. 1, pp. 294-315, 2015, doi:10.4271/2015-01-1557.

[13] P. Mousley, *Flyer Cobra Probe,* T. F. I. P. Ltd., Ed., Australia, 2016.

[14] E. Mercker and J. Wiedemann, "On the Correction of Interference Effects in Open Jet Wind Tunnels," *SAE Technical Paper 960671,* 1996, doi:10.4271/960671.

[15] E. Mercker and K. Cooper, "A Two-Measurement Correction for the Effects of a Pressure Gradient on Automotive, Open-Jet, Wind Tunnel Measurements," *SAE Technical Paper 2006-01-0568,* 2006, doi:10.4271/2006-01-0568.

[16] A. Hennig, Eine erweiterte Methode zur Korrektur von Interferenzeffekten in Freistrahlwindkanälen für Automobile, Doctoral Dissertation, Universität Stuttgart, Springer Vieweg, 2016, doi:10.1007/978-3-658-17827-7.

[17] T. Lounsberry and J. Walter, "Practical Implementation of the Two-Measurement Correction Method in Automotive Wind Tunnels," *SAE Int. J. Passeng. Cars - Mech. Syst.,* vol. 8, no. 2, pp. 676-686, 2015, doi:10.4271/2015-01-1530.

[18] O. Fischer, Investigation of Correction Methods for Interference Effects in Open-Jet Wind Tunnels, Doctoral Dissertation, Universität Stuttgart, Springer Vieweg, 2018, doi:10.1007/978-3-658-21379-4_4.

[19] A. Hennig and E. Mercker, "EADE Correlation Test Report 2010," EADE-Meeting at FIAT, Turin, 2011.

[20] T. Schütz, N. Grün und R. Blumrich, „Numerische Methoden," in *Hucho-Aerodynamik des Automobils: Strömungsmechanik, Wärmetechnik, Fahrdynamik, Komfort*, Springer-Verlag, 2013.

[21] R. Lietz, S. Mallick, S. Kandasamy and H. Chen, "Exterior Airflow Simulations Using a Lattice Boltzmann Approach," *SAE Technical Paper 2002-01-0596*, 2002, doi:10.4271/2002-01-0596.

[22] A. Heft, T. Indinger and N. Adams, "Introduction of a New Realistic Generic Car Model for Aerodynamic Investigations," *SAE Technical Paper 2012-01-0168*, 2012, doi:10.4271/2012-01-0168.

[23] C. Zhang, M. Tanneberger, T. Kuthada, F. Wittmeier, J. Wiedemann and J. Nies, "Introduction of the AeroSUV - A New Generic SUV Model for Aerodynamic Research," *SAE Technical Paper 2019-01-0646*, 2019, doi:10.4271/2019-01-0646.

[24] S. Chaligné, R. Turner and A. Gaylard, "The Aerodynamics Development of the New Land Rover Discovery," in *Progress in Vehicle Aerodynamics and Thermal Management*, Stuttgart, 2017.

[25] L. Larson, R. Gin and R. Lietz, "Aerodynamic Investigation of Cooling Drag of a Production Sedan Part 2: CFD Results," *SAE Int. J. Passeng. Cars - Mech. Syst.*, vol. 10, no. 1, pp. 214-223, 2017, doi:10.4271/2017-01-1528.

[26] J. Saunders und R. Mansour, „On-Road and Wind Tunnel Turbulence and its Measurement Using a Four-Hole Dynamic Probe Ahead of Several Cars," *SAE Technical Paper 2000-01-0350*, 2000, doi:10.4271/2000-01-0350.

[27] S. S. J. Watkins, "A Review of the Wind Conditions Experienced by a Moving Vehicle," *SAE Technical Paper 981182*, 1998, doi:10.4271/981182.

[28] K. Cooper and S. Watkins, "The Unsteady Wind Environment of Road Vehicles, Part One," *SAE Technical Paper 2007-01-1236,* 2007, doi:10.4271/2007-01-1236.

[29] S. Wordley and J. Saunders, "On-road Turbulence: Part 2," *SAE Int. J. Passeng. Cars - Mech. Syst.,* vol. 2, no. 1, pp. 111-137, 2009, doi:10.4271/2009-01-0002.

[30] N. Oettle, D. Sims-Williams, R. Dominy, C. Darlington, C. Freeman and P. Tindall, "The Effects of Unsteady On-Road Flow Conditions on Cabin Noise," *SAE Technical Paper 2010-01-0289,* 2010, doi:10.4271/2010-01-0289.

[31] D. Schröck, N. Widdecke and J. Wiedemann, "Aerodynamic Response of a Vehicle Model to Turbulent Wind," in *Proceedings of the 7th FKFS Conference, Progress in Vehicle Aerodynamics and Thermal Management,* Stuttgart, 2009.

[32] D. Schröck, N. Widdecke and J. Wiedemann, "On-Road Wind Conditions Experienced by a Moving Vehicle," in *Proceedings of the 6th FKFS Conference, Progress in Vehicle Aerodynamics,* Stuttgart, 2007.

[33] A. Wagner, Ein Verfahren zur Vorhersage und Bewertung der Fahrerreaktion bei Seitenwind, Doctoral Dissertation, Universität Stuttgart, expert Verlag, 2003.

[34] F. Buckley, C. Marks and W. Walston, "Study of aerodynamic methods for improving truck fuel economy," College Park (USA), 1978.

[35] K. Ingram, "The Wind Averaged Drag Coefficient Applied to Heavy Goods Vehicles," Transport and Road Research Laboratory, Wokingham, Berkshire, United Kingdom, 1978.

[36] J. Leuschen and K. Cooper, "Full-Scale Wind Tunnel Tests of Production and Prototype, Second-Generation Aerodynamic Drag-Reducing Devices for Tractor-Trailers," *SAE Technical Paper 2006-01-3456,* 2006, doi:10.4271/2006-01-3456.

[37] P. Gururaja, "Wind-Averaged Drag Determination for Heavy-Duty Vehicles Using On-Road Constant-Speed Torque Tests," *SAE Technical Paper 2016-01-8153,* 2016, doi:10.4271/2016-01-8153.

[38] S. Windsor, "Real world drag coefficient - is it wind averaged drag?," *International Vehicle Aerodynamic Conference,* pp. 3-18, October 2014, doi:10.1533/9780081002452.1.3.

[39] J. Howell, D. Forbes and M. Passmore, "A drag coefficient for application to the WLTP driving cycle," *Proceedings of the Institution of Mechanical Engineers, Part D: Journal of Automobile Engineering,* vol. 231, no. 9, pp. 1274-1286, 2017, doi:10.1177/0954407017704784.

[40] D. Stoll, Ein Beitrag zur Untersuchung der aerodynamischen Eigenschaften von Fahrzeugen unter böigem Seitenwind, Doctoral Dissertation, Universität Stuttgart, Springer Fachmedien Wiesbaden, 2018, doi:10.1007/978-3-658-21545-3.

[41] S. Watkins, M. Riegel and J. Wiedemann, "The effect of turbulence on wind noise: a road and wind-tunnel study," in *Proceedings, 5. Internationales Stuttgarter Symposium,* 2003.

[42] S. Krampol, M. Riegel and J. Wiedemann, "Noise synthesis - a procedure to simulate the turbulent noise interior of cars," in *Proceedings of the 7th FKFS-Conference, Progress in Vehicle Aerodynamics and Thermal Management,* Stuttgart, 2010.

[43] S. Watkins, Wind-Tunnel Modelling of Vehicle Aerodynamics: with Emphasis on Turbulent Wind Effects on Commercial Vehicle drag, RMIT University, 1990.

[44] A. Cogotti, "Ground Effect of a Simplified Car Model in Side-Wind and Turbulent Flow," *SAE Technical Paper 1999-01-0652,* 1999, doi:10.4271/1999-01-0652.

[45] D. Schröck, N. Widdecke and J. Wiedemann, "The Effect of High Turbulence Intensities on Surface Pressure Fluctuations and Wake Structures of a Vehicle Model," *SAE Int. J. Passeng. Cars - Mech. Syst.,* vol. 2, no. 1, pp. 98-110, 2009, doi:10.4271/2009-01-0001.

[46] A. Ryan and R. Dominy, "The Aerodynamic Forces Induced on a Passenger Vehicle in Response to a Transient Cross-Wind Gust at a Relative Incidence of 30°," *SAE Technical Paper 980392,* 1998, doi:10.4271/980392.

[47] R. Dominy and A. Ryan, "An Improved Wind Tunnel Configuration for the Investigation of Aerodynamic Cross Wind Gust Response," *SAE Technical Paper 1999-01-0808,* 1999, doi:10.4271/1999-01-0808.

[48] S. Mullarkey, Aerodynamic stability of road vehicles in side winds and gusts, Doctoral Dissertation, Imperial College London, London, 1990.

[49] M. Passmore, S. Richardson and A. Imam, "An experimental study of unsteady vehicle aerodynamics," *Proceedings of the Institution of Mechanical Engineers, Part D: Journal of Automobile Engineering,* vol. 215, no. 7, pp. 779-788, 2001.

[50] O. Mankowski, D. Sims-Williams und R. Dominy, „A Wind Tunnel Simulation Facility for On-Road Transients," *SAE Int. J. Passeng. Cars - Mech. Syst.,* Bd. 7, Nr. 3, pp. 1087-1095, 2014, doi:10.4271/2014-01-0587.

[51] T. Yamashita, T. Makihara, K. Maeda and K. Tadakuma, "Unsteady Aerodynamic Response of a Vehicle by Natural Wind Generator of a Full-Scale Wind Tunnel," *SAE Int. J. Passeng. Cars - Mech. Syst.,* vol. 10, no. 1, pp. 358-368, 2017, doi:10.4271/2017-01-1549.

[52] D. Schröck, Eine Methode zur Bestimmung der aerodynamischen Eigenschaften eines Fahrzeugs unter böigem Seitenwind, Doctoral Dissertation, Universität Stuttgart, Stuttgart: expert Verlag, 2011.

[53] F. Wittmeier, "The Recent Upgrade of the Model Scale Wind Tunnel of University of Stuttgart," *SAE Int. J. Passeng. Cars - Mech. Syst.,* vol. 10, no. 1, pp. 203-213, 2017, doi:10.4271/2017-01-1527.

[54] A. Michelbach and R. Blumrich, "Upgrade of the full-scale aeroacoustic wind tunnel of Stuttgart University by FKFS," *15. Internationales Stuttgarter Symposium. Proceedings,* May 2015, doi:10.1007/978-3-658-08844-6_33.

[55] A. Cogotti, "Update on the Pininfarina "Turbulence Generation System" and its effects on the car Aerodynamics and Aeroacoustics," *SAE Technical Paper 2004-01-0807,* 2004, doi:10.4271/2004-01-0807.

[56] G. C. A. Carlino, "Simulation of Transient Phenomena with the Turbulence Generation System in the Pininfarina Wind Tunnel," *SAE Technical Paper 2006-01-1031,* 2006, doi:10.4271/2006-01-1031.

[57] H. Wilhelmi, C. Jessing, J. Bell, D. Heine, A. Wagner, J. Wiedemann and C. Wagner, "Simulation of Transient On-Road Conditions in a Closed Test Section Wind Tunnel Using a Wing System with Active Flaps," *SAE Int. J. Adv. & Curr. Prac. in Mobility,* vol. 2, no. 5, pp. 2604-2616, 2020, doi:10.4271/2020-01-0688.

[58] D. Tang, P. G. Cizmas and E. Dowell, "Experiments and Analysis for a Gust Generator in a Wind Tunnel," *Journal of Aircraft,* vol. 33, no. 1, pp. 139-148, 1996, doi:10.2514/3.46914.

[59] A. Gaylard, N. Oettle, J. Gargoloff and B. Duncan, "Evaluation of Non-Uniform Upstream Flow Effects on Vehicle Aerodynamics," *SAE Int. J. Passeng. Cars - Mech. Syst.,* vol. 7, no. 2, pp. 692-702, 2014, doi:10.4271/2014-01-0614.

[60] A. D'Hooge, R. Palin, L. Rebbeck, J. Gargoloff and B. Duncan, "Alternative Simulation Methods for Assessing Aerodynamic Drag in Realistic Crosswind," *SAE Int. J. Passeng. Cars - Mech. Syst.,* vol. 7, no. 2, pp. 617-625, 2014, doi:10.4271/2014-01-0599.

[61] Dassault Systemes Simulia Corp., *PowerFLOW Best Practices Guide for External Aerodynamics, PF2020-R3-BP,* 2020.

[62] B. Duncan, L. D'Alessio, J. Gargoloff and A. Alajbegovic, "Vehicle aerodynamics impact of on-road turbulence," *Proceedings of the Institution of Mechanical Engineers, Part D: Journal of Automobile Engineering,* vol. 231, no. 9, pp. 1148-1159, 2017, doi:10.1177/0954407017699710.

[63] J. Gargoloff, B. Duncan, E. Tate, A. Alajbegovic, A. Belanger and B. Paul, "Robust Optimization for Real World CO2 Reduction," *SAE Technical Paper 2018-37-0015,* 2018, doi:10.4271/2018-37-0015.

[64] D. Stoll, T. Kuthada and J. Wiedemann, "Experimental and numerical investigation of aerodynamic drag in turbulent flow conditions," *IMechE International Conference on Vehicle Aerodynamics,* pp. 21-22, 2016.

[65] O. Fischer, T. Kuthada, N. Widdecke and J. Wiedemann, "CFD Investigations of Wind Tunnel Interference Effects," *SAE Technical Paper 2007-01-1045,* 2007, doi:10.4271/2007-01-1045.

[66] O. Fischer, T. Kuthada, J. Wiedemann, P. Dethioux, R. Mann and B. Duncan, "CFD Validation Study for a Sedan Scale Model in an Open Jet Wind Tunnel," *SAE Technical Paper 2008-01-0325,* 2008, doi:10.4271/2008-01-0325.

[67] O. Fischer, T. Kuthada, E. Mercker, J. Wiedemann and B. Duncan, "CFD Approach to Evaluate Wind-Tunnel and Model Setup Effects on Aerodynamic Drag and Lift for Detailed Vehicles," *SAE Technical Paper 2010-01-0760,* 2010, doi:10.4271/2010-01-0760.

[68] D. Stoll, T. Kuthada and J. Wiedemann, "Unsteady Aerodynamic Vehicle Properties of the DrivAer Model in the IVK Model Scale Wind Tunnel," in *Proceedings of the 10th FKFS Conference, Progress in Vehicle Aerodynamics and Thermal Management,* Stuttgart, 2015.

[69] X. Fei, C. Jessing, T. Kuthada, J. Wiedemann and A. Wagner, "The Influence of Different Unsteady Incident Flow Environments on Drag Measurements in an Open Jet Wind Tunnel," *Fluids,* vol. 5, no. 4, 2020, doi:10.3390/fluids5040178.

[70] C. Jessing, H. Wilhelmi, F. Wittmeier and J. Wiedemann, "Characterization of the Transient Airflow Around a Vehicle on Public Highways," in *Proceedings of the 12th FKFS Conference, Progress in Vehicle Aerodynamics and Thermal Management*, Stuttgart, 2019.

[71] SAE Road Vehicle Aerodynamics Committee, "Aerodynamic Testing of Road Vehicles - Open Throat Wind Tunnel Adjustment," 1994.

[72] T. Kuthada, E. Pfannkuchen and J. Wiedemann, "Evaluation of Cooling Air Drag on a Reference Body," in *5th MIRA International Vehicle Aerodynamics Conference*, Warwick, UK, 2004.

[73] F. Wittmeier and T. Kuthada, "Open Grille DrivAer Model - First Results," *SAE Int. J. Passeng. Cars - Mech. Syst.,* vol. 8, no. 1, pp. 252-260, 2015, doi:10.4271/2015-01-1553.

[74] X. Fei, T. Kuthada, A. Wagner and J. Wiedemann, "The Effect of Unsteady Incident Flow on Drag Measurements for Different Vehicle Geometries in an Open Jet Wind Tunnel," *SAE Int. J. Adv. & Curr. Prac. in Mobility,* vol. 4, no. 6, pp. 1999-2011, 2022, doi:10.4271/2022-01-0894.

[75] J. Wiedemann, O. Fischer and P. Jiabin, "Further Investigations on Gradient Effects," *SAE Technical Paper 2004-01-0670,* 2004, doi:10.4271/2004-01-0670.

[76] K. Cooper, E. Mercker and J. Müller, "The necessity for boundary corrections in a standard practice for the open-jet wind tunnel testing of automobiles," in *IMechE J. of Automobile Engineering, Automotive Aerodynamics Special Issue*, 2017.

[77] B. Bock, Potentialabschätzung zur Luftwiderstandsreduktion am stumpfen Körper durch Maßnahmen der Strömungsbeeinflussung, Doctoral Dissertation, Universität Stuttgart, Stuttgart, 2019, doi:10.18419/opus-10722.

[78] C. Schönleber, Untersuchung von transienten Interferenzeffekten in einem Freistrahlwindkanal für Automobile, Doctoral Dissertation, Universität Stuttgart, Stuttgart: Springer Vieweg, 2020, doi:10.1007/978-3-658-32718-7.

Appendix

A1. Incident Flow Signals Used

All investigations were performed in 1:4 scale environments. Signal turbulence values can be calculated as is (in 1:4 scale), or converted to 1:1 scale using dimensionless quantities to facilitate comparisons with 1:1 scale research.

Turbulence intensity (Tu) and length scale (TLS) are calculated using the following values:

- Time-resolved flow angle $\beta(t)$
- Flow velocity v
- Sampling frequency f_s

Of these values,

- Flow angle has no dimensionless quantity associated with it that affects it during model scale investigations.
- Flow velocity is usually higher in model scale to account for Reynold number equivalency. During measurements and simulations in this work, nominal flow velocity was 45 m/s (162 km/h). For the comparison, a velocity of 100 km/h was chosen because it's the velocity used in Wordley and Saunders' investigations [7]. It also lies within Jessing's [70] measurement range of 80–140 km/h.
- The sampling frequency needs to be adjusted to achieve the same Strouhal number in 1:1 and 1:4 scales, at 100 and 162 km/h respectively. The formula can be found in Eq. 2.8.

Table 6.1 and **Table 6.2** show all signal characteristic values, measured in the empty MWK test section, Tu_y and TLS_y are shown both in original 1:4 scale at 162 km/h as well as in converted 1:1 scale at 100 km/h.

The noise signals' standard deviations and turbulence values (1:1 scale) fall in the following ranges (comparison with on-road measurements from Jessing [4] in parentheses):

© The Editor(s) (if applicable) and The Author(s), under exclusive license to
Springer Fachmedien Wiesbaden GmbH, part of Springer Nature 2026
X. Fei, *The Impact of Unsteady Flow on Drag Measurements in Automotive
Wind Tunnels*, Wissenschaftliche Reihe Fahrzeugtechnik Universität Stuttgart,
https://doi.org/10.1007/978-3-658-51766-3

- σ: 1.1–3.9° (2.2–2.9°)
- Tu_y: 0.9–5.8 % (3–13 %)
- TLS_y: 0.27–1.14 m (0.25–4.5 m)

The values are therefore largely in agreement with Jessing's results, which are some of the most recent comprehensive on-road measurements made.

Table 6.1:   Filtered Random Noise Signals

f_c in Hz	St_c	σ_{flap} in deg	σ in deg	Tu_y in % (1:4)	TLS_y in m (1:4)	Tu_y in % (1:1)	TLS_y in m (1:1)
1.5	0.038	0.75	1.06	1.61	0.194	0.92	0.278
1.5	0.038	1.50	1.97	2.88	0.290	1.70	0.640
1.5	0.038	2.25	2.95	4.30	0.498	2.42	0.911
1.5	0.038	3.00	3.67	5.56	0.713	2.97	1.086
3.0	0.077	0.75	1.08	1.80	0.149	1.23	0.343
3.0	0.077	1.50	2.04	3.36	0.204	2.57	0.639
3.0	0.077	2.25	2.95	4.78	0.322	3.55	0.927
3.0	0.077	3.00	3.84	6.14	0.426	4.52	1.140
5.0	0.128	0.75	1.15	1.92	0.084	1.58	0.328
5.0	0.128	1.50	2.16	3.63	0.164	2.89	0.590
5.0	0.128	2.25	3.12	5.21	0.228	4.21	0.821
5.0	0.128	3.00	3.90	6.51	0.321	5.15	0.963
7.5	0.192	0.75	1.08	1.85	0.062	1.60	0.272
7.5	0.192	1.50	2.09	3.57	0.104	3.11	0.503
7.5	0.192	2.25	2.99	5.12	0.138	4.46	0.734
7.5	0.192	3.00	3.82	6.51	0.184	5.77	0.853

Table 6.2:   Sine Signals (*italics: used in CFD investigations*)

f in Hz	St	A_{flap} in deg	A in deg	Tu_y in % (1:4)	TLS_y in m (1:4)	Tu_y in % (1:1)	TLS_y in m (1:1)
1.5	0.038	1.20	1.23	1.51	0.064	1.51	0.350
1.5	0.038	2.50	2.48	3.06	0.133	3.05	0.740
1.5	*0.038*	*5.00*	*4.44*	*5.46*	*0.231*	*5.45*	*1.280*
1.5	0.038	7.50	5.98	7.32	0.298	7.31	1.633
1.5	0.038	10.00	7.53	9.27	0.374	9.28	2.057
3.0	0.077	1.20	0.90	1.12	0.023	1.11	0.158
3.0	*0.077*	*2.50*	*2.07*	*2.56*	*0.056*	*2.55*	*0.374*
3.0	*0.077*	*5.00*	*4.17*	*5.12*	*0.108*	*5.10*	*0.713*
3.0	0.077	7.50	5.80	7.08	0.143	7.06	0.919
3.0	0.077	10.00	7.25	8.89	0.179	8.89	1.126
5.0	0.128	1.20	0.84	1.03	0.013	1.03	0.082
5.0	0.128	2.50	1.88	2.32	0.030	2.32	0.193
5.0	*0.128*	*5.00*	*3.94*	*4.84*	*0.061*	*4.82*	*0.386*
5.0	0.128	7.50	6.01	7.41	0.089	7.39	0.548
5.0	0.128	10.00	7.52	9.23	0.111	9.20	0.677
7.5	0.192	1.20	0.92	1.14	0.010	1.13	0.062
7.5	*0.192*	*2.50*	*1.98*	*2.44*	*0.021*	*2.44*	*0.135*
7.5	*0.192*	*5.00*	*4.11*	*5.06*	*0.042*	*5.04*	*0.266*
7.5	0.192	7.50	6.34	7.81	0.063	7.81	0.384
7.5	0.192	10.00	8.07	9.94	0.080	9.94	0.489

A2. Pressure Distribution Self-Similarity

Figure 6.1 shows three static pressure distributions for the same incident flow signal in the Model Scale Wind Tunnel. They differ in the wind tunnel collector setup:

- "none": no stagnation bodies in collector
- "S": stagnation bodies at 48 mm extension
- "L": stagnation bodies at 76 mm extension

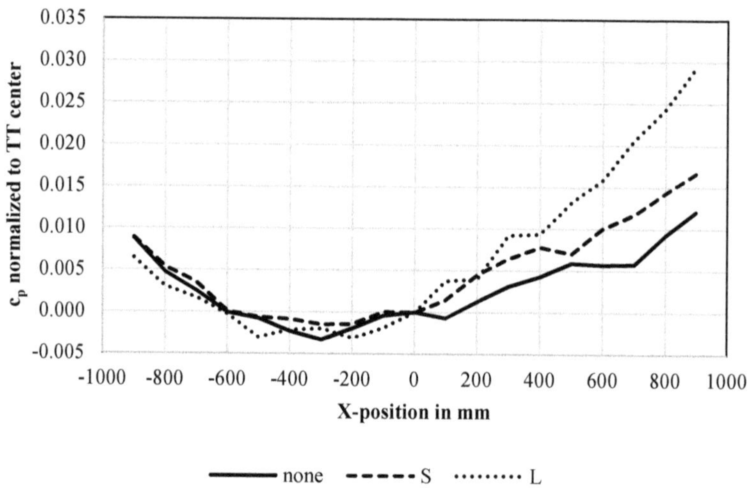

Figure 6.1:   Static pressure distributions in the MWK for one incident flow signal but different stagnation bodies (sine signal, f = 7.5 Hz, $A_{nominal}$ = 7.5°)

These three pressure distributions are not self-similar to each other. It is impossible to convert one pressure distribution into the other by moving it horizontally along the x-axis and then multiplying all its values with a constant factor.

Figure 6.2 shows the same pressure distribution curve for "none" as **Figure 6.1**. However, the other two curves are the "none" curve multiplied by a constant factor so that their c_p values at x = 500 mm are equal to those of "48 mm" and "76 mm" in **Figure 6.1**. Thus, the curves in **Figure 6.2** are self-similar.

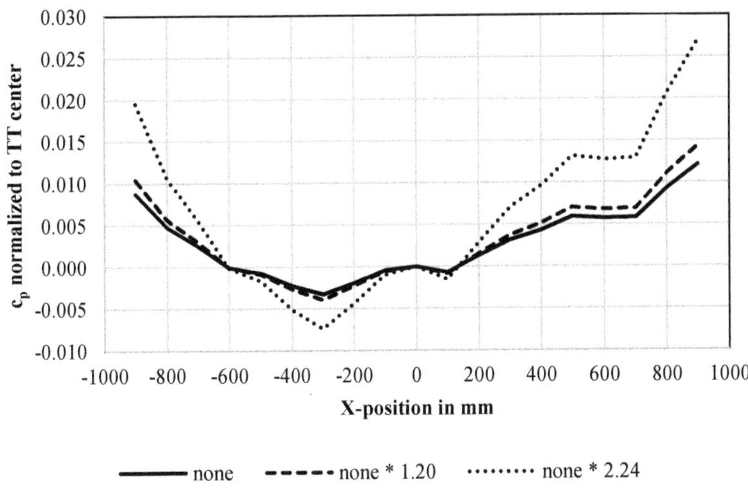

Figure 6.2: Example of three self-similar pressure distributions, with the "none" pressure distribution from **Figure 6.1** used as a baseline

While this has an insignificant impact on accuracy during steady-state testing, the greatly increased static pressure near the collector during unsteady testing (see section 4.2.1) will compound the issue. Higher gradients near the collector means higher pressure deltas from front to back in the test section, increasing correction magnitudes and amplify existing correction inaccuracies.

Figure 6.3 shows a comparison between pressure distributions under steady and unsteady flow, where the unsteady flow pressure distributions are the same as in **Figure 6.1**. It is clearly visible that the static pressure near the collector is much larger in unsteady flow than in steady flow, even with the largest (76 mm) stagnation bodies installed for the steady flow case. This proves that pressure distribution self-similarity is a bigger concern in unsteady flow environments.

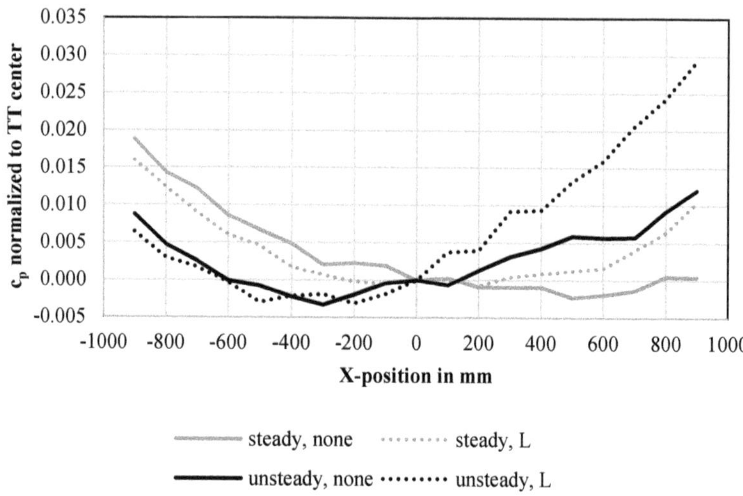

Figure 6.3:   Comparison of pressure distributions in steady and unsteady flow, with and without stagnation bodies (unsteady flow as in **Figure 6.1**)

A3. Corrected c_D plots over standard deviation σ

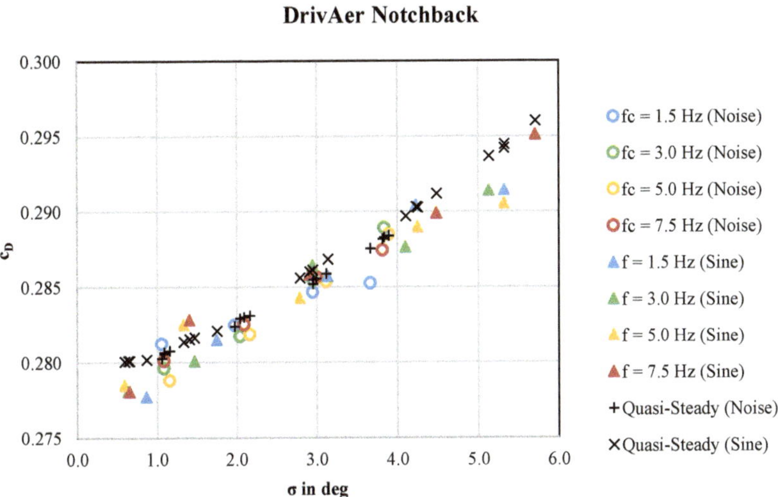

Figure 6.4: DrivAer notchback

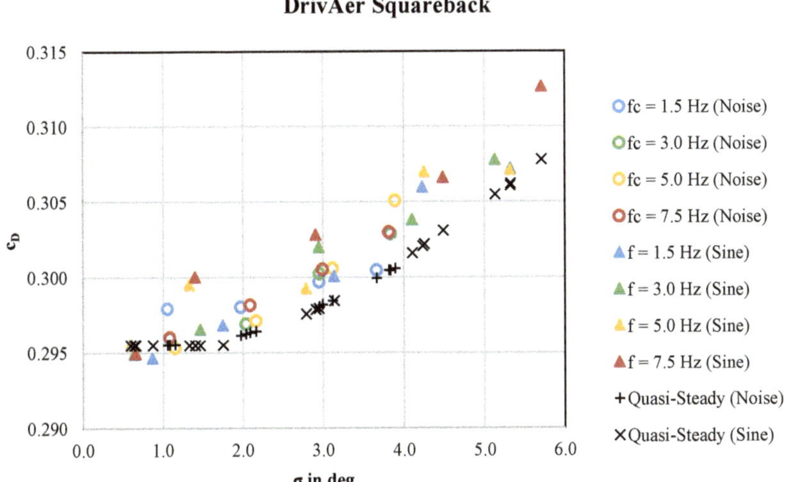

Figure 6.5: DrivAer squareback

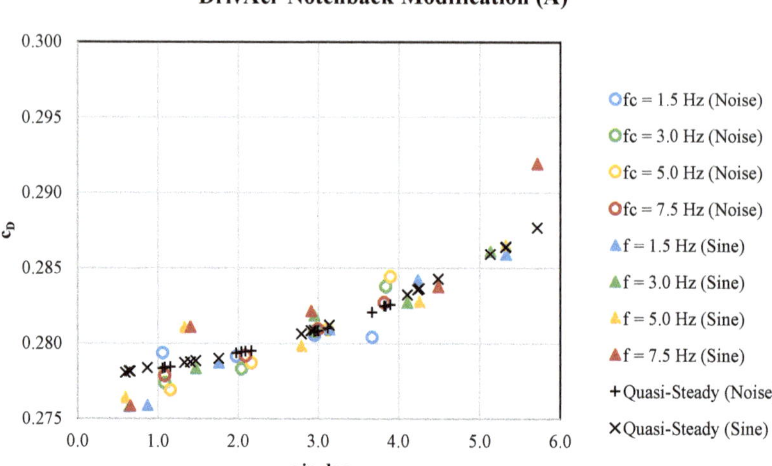

Figure 6.6:   DrivAer notchback with modification A

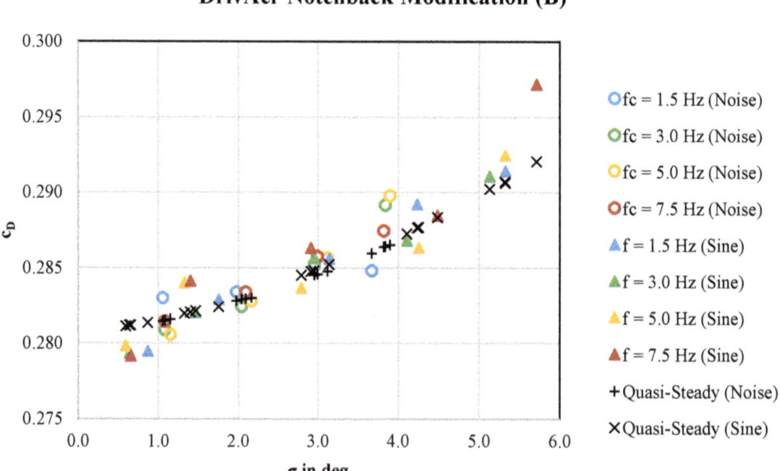

Figure 6.7:   DrivAer notchback with modification B

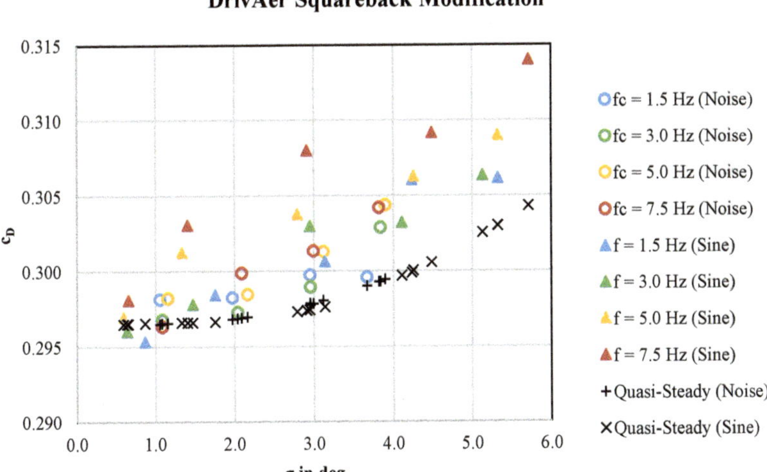

Figure 6.8: DrivAer squareback with modification

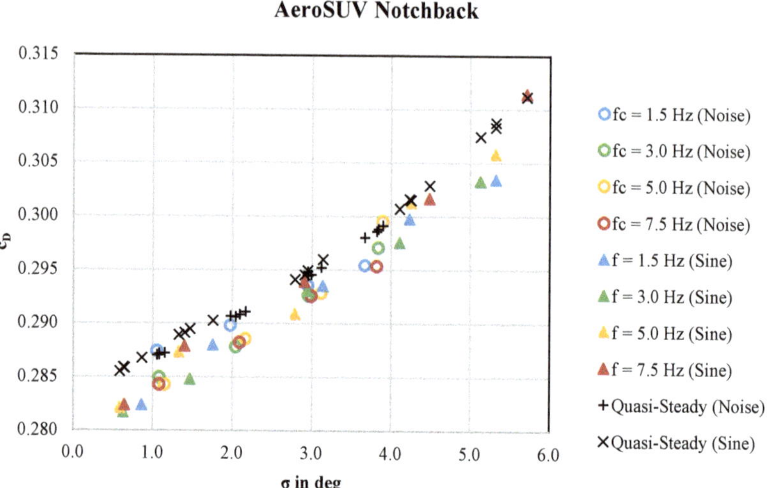

Figure 6.9:   AeroSUV notchback

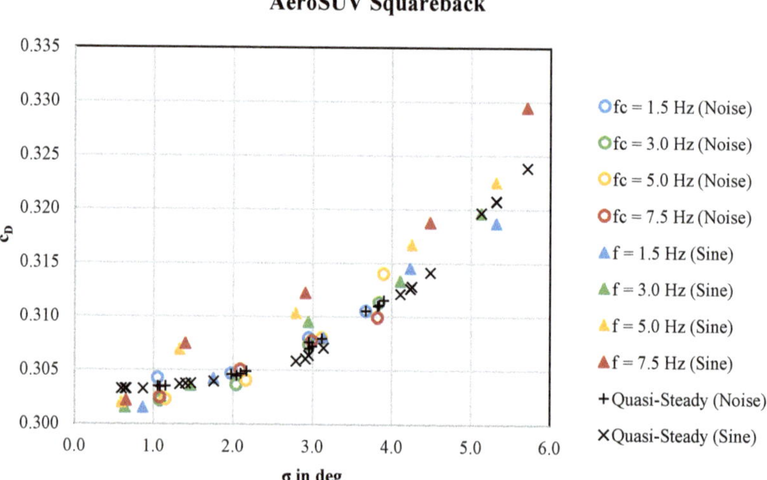

Figure 6.10:   AeroSUV squareback

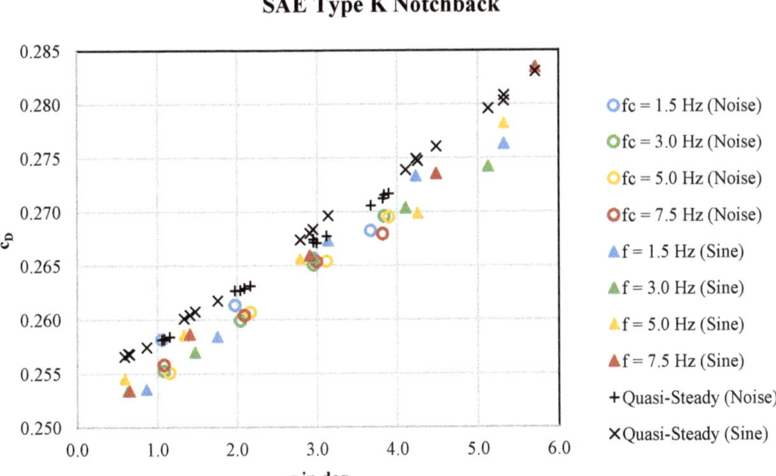

Figure 6.11:   SAE Type K notchback

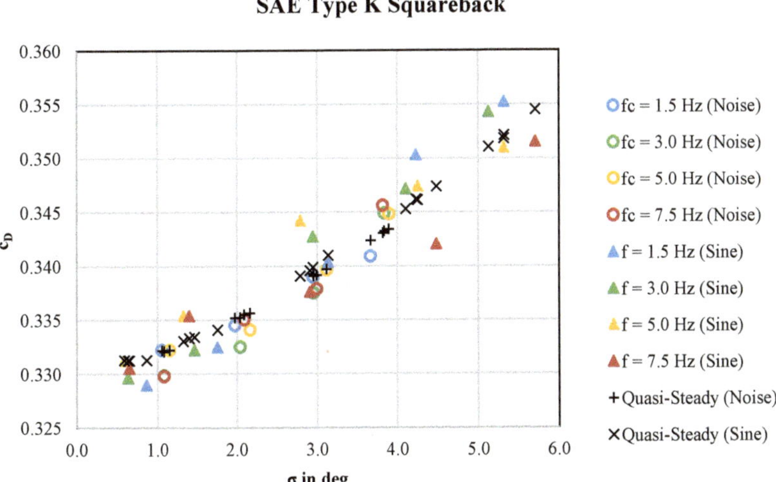

Figure 6.12:   SAE Type K squareback

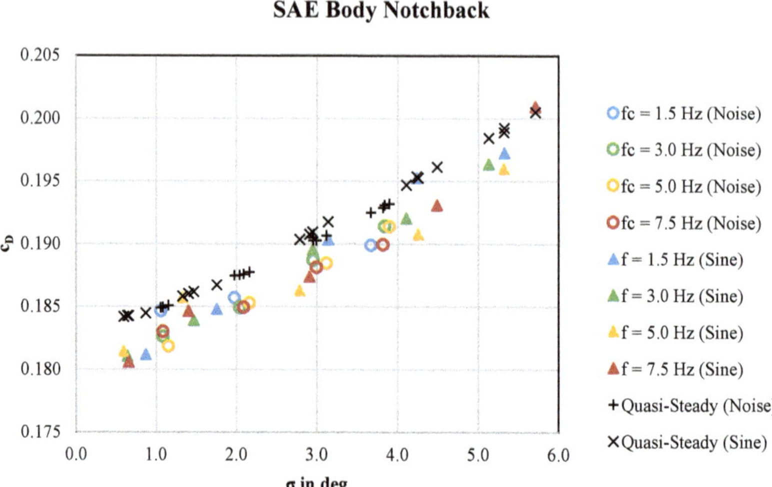

Figure 6.13: SAE Body notchback

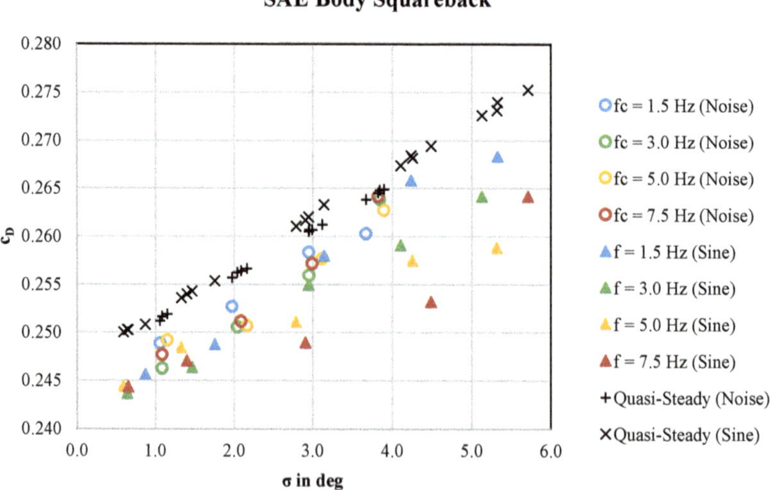

Figure 6.14: SAE Body squareback

GPSR Compliance
The European Union's (EU) General Product Safety Regulation (GPSR) is a set
of rules that requires consumer products to be safe and our obligations to
ensure this.

If you have any concerns about our products, you can contact us on

ProductSafety@springernature.com

In case Publisher is established outside the EU, the EU authorized
representative is:

Springer Nature Customer Service Center GmbH
Europaplatz 3
69115 Heidelberg, Germany

www.ingramcontent.com/pod-product-compliance
Ingram Content Group UK Ltd.
Pitfield, Milton Keynes, MK11 3LW, UK
UKHW020230080726
473059UK00001B/11